D0311855

PROBLEMS IN CONSTRUCTION CLAIMS

Also by Vincent Powell-Smith and published by BSP Professional Books

A Building Contract Casebook
Second Edition
jointly with Michael Furmston

Building Contract Claims
Second Edition
jointly with John Sims

The Building Regulations Explained and Illustrated
Eighth Edition
jointly with M.J. Billington

Civil Engineering Claims
jointly with Douglas Stephenson

Contract Documentation for Contractors
Second Edition
jointly with John Sims

Determination and Suspension of Construction Contracts
jointly with John Sims

GC/Works/1
Edition 3

Problems in Construction Claims

Some Building Contract Problems

PROBLEMS IN CONSTRUCTION CLAIMS

VINCENT POWELL-SMITH

LLM, DLitt, FCIArb, MBAE

OXFORD

BSP PROFESSIONAL BOOKS

LONDON EDINBURGH BOSTON

MELBOURNE PARIS BERLIN VIENNA

First published 1990

British Library
Cataloging in Publication Data
Powell-Smith, Vincent
Problems in construction claims.
1. Great Britain. Buildings. Construction.
Contracts. Claims
I. Title
344.1037869

ISBN 0-632-02852-1

BSP Professional Books
A division of Blackwell Scientific
Publications Ltd
Editorial Offices:
Osney Mead, Oxford OX2 0EL
(Orders: Tel. 0865 240201)
25 John Street, London WC1N 2BL
23 Ainslie Place, Edinburgh EH3 6AJ
3 Cambridge Center, Suite 208, Cambridge
MA 02142, USA
54 University Street, Carlton, Victoria 3053,
Australia

Set by DP Photosetting, Aylesbury, Bucks
Printed and bound in Great Britain by
Billing & Sons Ltd, Worcester

Dedication

For John Haesaert Mancel Sims
– who does not need it!

Contents

Preface

Considerable time and energy is devoted to the making and settling of money claims under the various standard forms of contract in use in the construction industry, all of which contain provisions enabling the contractor to claim reimbursement under the contract for the cost of disruption or prolongation. In some cases, it takes longer to prepare and agree the final account than it does to construct the building.

The subject of 'claims' is an emotive one and the basis of the contractual provisions is often misunderstood. Those responsible for administering the contract and for actually ascertaining and settling claims, as well as contractors and sub-contractors, frequently fail to grasp the legal principles involved, as the growing body of case law shows.

This short book deals with some of the problems faced by contractors, sub-contractors, architects, engineers and quantity surveyors in the making and settling of prolongation and disruption claims. It also looks at the related problems of extensions of time and liquidated and ascertained damages. Its content is based not only on case law but also on practice – and my professional experience of both over the last twenty-years shows that many of the problems recur time and again. It is not a complete coverage of every aspect of claims procedure and practice. Those who wish for a full treatment of all facets of the subject should refer to *Building Contract Claims* (2nd edition, 1988, BSP) which I wrote with John Sims.

Here, I have concentrated on some of the more common problems illustrated by the cases. The coverage is wide and extends to most of the standard building and civil engineering contracts in common use.

It is a book for practical people and will, I hope, be treated as a contribution to the understanding of a complex and difficult area. It represents *one* construction industry lawyer's view of areas of common difficulty and, if nothing else, perhaps it will stimulate debate. Edition 3 of GC/Works/1 was published just as the book was being finished and so I have touched briefly on some of the 'claims' aspects of that important document.

This book has been put together in the intervals allowed by a busy professional life involving much travel, and so I am doubly grateful to Julia Burden of the publishers for her help in seeing the book through the press at a time when I was difficult to contact as well as for our almost daily exchanges of facsimile messages! It is always a pleasure to work with Julia.

I am also grateful to Dr John Parris for his helpful comments and to the many friends and colleagues who over the years have contributed to my understanding of the subject, as well as to the countless delegates at many conferences who have asked me (and others) searching questions, the consideration of which has often led me to revise my opinions.

The law is stated as at December 1989.

Funchais
Portugal

Vincent Powell-Smith

List of Abbreviations

ACA 2	Association of Consultant Architects Form of Building Agreement, Second Edition, 1984
BEC	Building Employers Confederation
BPF	British Property Federation
CCC/Works/1	Government Conditions of Contract for Building and Civil Engineering, 1943–1959
CESMM	Civil Engineering Standard Method of Measurement
DOM/1	Standard Form of Domestic Sub-contract for use with JCT 80
FASS	Federation of Associations of Specialists and Sub-contractors
FCEC	Federation of Civil Engineering Contractors
FIDIC	Fédération Internationale des Ingénieurs Conseils
GC/Works/1	Government General Conditions of Contract for Building and Civil Engineering, 2nd edition, 1977, 3rd edition, 1989
ICE Conditions	Institution of Civil Engineers Conditions of Contract, 5th Edition
ICE Minor Works Conditions	Institution of Civil Engineers Conditions of Contract for Minor Works 1988
IFC 84	JCT Intermediate Form of Building Contract 1984
JCT	Joint Contracts Tribunal for the Standard Form of Building Contract
JCT 63	JCT Standard Form of Building Contract 1963
JCT 80	JCT Standard Form of Building Contract 1980
NAM/SC	Named Sub-contractors Conditions for use with JCT Intermediate Form 1984
NSC/T	JCT Nominated Sub-contract Tender and Agreement for use with JCT 80

NFBTE	National Federation of Building Trades Employers
NFBTE/FASS 'Green Form'	Standard Form of Sub-contract for Nominated Sub-contractors for use with JCT 63
NFBTE/FASS 'Blue Form'	Standard Form of (Non-nominated) Sub-contract for use with JCT 63
NW80	JCT Agreement for Minor Building Works 1980
PSA	Property Services Agency
RIBA Form	Royal Institute of British Architects Standard Form of Building Contract 1939
SMM7	Standard Method of Measurement for Building Works, 7th Edition

Chapter 1

Background: Some Principles of Contractual Claims

Introduction

The majority of construction work is carried out on standard form terms and two lines or families of building contract forms have evolved in the UK. The JCT Forms have evolved from a standard contract originating in the last century. Their recognisable first predecessor was issued in 1909 and they are the outcome of mutual agreement between representatives of two sides of the intended contracts. Being negotiated, they are compromise or consensus documents. For present purposes, JCT 80 and IFC 84 may be taken as typical of this family. JCT 80 was a vast improvement on its predecessor, JCT 63, but most of the previous text was maintained and the significant changes were mainly procedural rather than legal in effect. The case law now developing in the courts largely concerns JCT 63 – at least so far as money claims are concerned – but for the most part is of application to JCT 80 and the other JCT forms where similar, if not identical, wording is used.

The second line of forms is that sponsored by the government, of which the main example is GC/Works/1, which dates from 1973, its Edition 2 being published in September 1977, as amended to 17 November 1977. In December 1989 the PSA – which is the main user of the form – published Edition 3 of GC/Works/1 which contains many radical new provisions and is effectively a completely new form.

GC/Works/1 also has a respectable history since the version used in 1959 (and then called CCC/Works/1) was noted as being 'Edition 9'. GC/Works/1 is drafted principally with the employer's interests in mind and is more draconian than the middle-of-the-way JCT forms. It is a form of unilateral provenance, though in modern practice the contracting side of the industry is consulted about intended changes.

The majority of civil engineering work is carried out under the ICE Conditions, 5th edition, published by the Institution of Civil Engineers, and which is also a negotiated contract with a long history. More recently, the ICE Conditions of Contract for Minor Works were published and are becoming widely used.

All these contracts provide specific machinery for the making and settlement of money claims and the grant of extensions of time, and as will be seen there is not any necessary link between the two.

Claims

It is important to realise that there are two substantially different types of claim which may be made in contract:

(a) *Claims for breach of contract* (sometimes called 'common law claims') when the claimant needs to prove a breach of contract and is then entitled to recover damages calculated on common law principles. These claims must be pursued in arbitration or litigation.

(b) *Claims under the contract.* These arise because some provision in the contract entitles the contractor to payment for 'loss' or 'expense', and are made and settled under machinery provided by the contract itself. In some cases, events which give rise to such a claim will also be breaches of contract and give rise to a common law claim. In other cases they will not, e.g., the issue of a variation order may give rise to a loss or expense claim even though it is authorised by the contract itself.

Many of the principles governing the permissible extent of a claim in contract are the same whether it is under the standard form itself or for breach of contract at common law: see *F.G. Minter Ltd* v. *Welsh Health Technical Services Organisation* (1980). In both cases the burden of proof is on the claimant, and it is for him to substantiate his claim. The principal benefit to the contractor of a claim under the contract (even if the event relied on also amounts to a breach of contract) is that its amount is quantified and paid under the contractual machinery as work proceeds. The correct operation of the contractual claims provisions creates a right under the contract to a *debt* rather than a claim for damages, and a debt is enforceable by summary procedure.

Disruption and prolongation claims

There is a distinction between disruption and prolongation claims. Under the JCT contracts there is no separate provision for the recovery of prolongation or disruption costs as such. Claims for the consequences of disruption or prolongation may, however, form part of the recovery of 'direct loss and/or expense'. GC/Works/1 Edition 2, clause 53, (Edition 3, clause 46) refers specifically to the recovery of prolongation and disruption expenses, thus emphasising that there are basically two claim situations:

- Delay in completion of the contract works beyond the date when they otherwise would have been completed. This is a *prolongation claim* – sometimes and inaccurately called a claim for extended preliminaries.
- A claim for the effect of an event upon the contract works themselves which does not necessarily involve a delay in completion of the works. This is a *disruption claim* and can arise even where the works are completed within the contract period. It is sometimes argued that disruption claims are not permissible under JCT terms because of the wording of the relevant clauses, e.g., JCT 80, clause 26.1 refers to 'the regular progress of the Works or any part thereof [being] materially affected by any one or more of' the matters listed in clause 26.2, which, it is said, means that it is only delay in progress which can be compensated and that the effect upon progress is to be equated with 'delay', thus ruling out any claim for loss of productivity. This view is erroneous. What matters is the effect of the stated event on *the regular progress of the works,* i.e., any delay to or disruption of the contract progress: the arguments are fully discussed by Powell-Smith and Sims in *Building Contract Claims,* 2nd edition, 1988, pp 168–71. Neither the JCT contracts nor GC/Works/1 requires a money claim either to be preceded by or accompanied by the grant of an extension of time, and to argue otherwise is a misconception.

Extensions of time and money claims

The questions of reimbursement of 'direct loss and/or expense' (JCT) or 'expense in performing the contract' (GC/Works/1) and the grant of extensions of time are quite separate and distinct. The purposes of the money claims provisions and those for extensions of time are different. The grant of an extension of time (GC/Works/1, clause 28(2)) (JCT 80, clause 25; IFC 84, clause 2.3) merely entitles the contractor to relief from paying liquidated damages from the date stated in the contract. He is not automatically entitled to any compensation because the supervising officer or architect has determined an extension of time: some of the events giving rise to an entitlement to an extension of time are neutral events in the sense that they are the fault or responsibility of neither party, e.g., the weather.

The supervising officer or architect can grant an extension of time without giving any 'loss' or 'expense' to the contractor. GC/Works/1 and IFC 84 emphasise the lack of connection between extensions of time and monetary claims by widely separating the provisions; JCT 80 puts the two provisions sequentially and seems to compound the fallacy by the provision in clause 26.3 which provides that, if and to the extent that it is necessary for the purpose of ascertainment of direct loss and/or expense, the architect must state in writing what extension of time, if any, he has granted under clause 25 in respect of those events which are also, independently, grounds for reimbursement under clause 26.

The position was clarified by the important case of *H. Fairweather & Co. Ltd* v. *London Borough of Wandsworth* (1987), which arose under a contract in JCT 63 form. The High Court ruled expressly that the grant of an extension of time under clause 23 is not a condition precedent to an application by a contractor for reimbursement of direct loss and/or expense under clause 24. This principle applies equally to JCT 80 and IFC 84. In fairness it must be said that the learned judge recognised that there was a practical connection between the grant of an extension of time under clause 23 (f) and the possibility of recovery of direct loss and/or expense under clause 24(1)(a).

Judge Fox-Andrews gave an example to illustrate that there is no necessary link between time and money. A strike occurs during a period of extension of time granted as a result of a variation order. The strike deprives the contractor of the opportunity to protect his machinery during the winter period when, in the example, he could not have carried out work, and so it takes him longer to complete the work. If the architect had granted an extension of time of only eight months under clause 24(d) on account of the strike, the contractor could still recover all his direct loss and/or expense under clause 11(6), which would include the loss arising as a result of the strike.

But two qualifications must be made:

- The direct loss and/or expense provisions only apply if the contractor is not otherwise adequately reimbursed under the contract.
- The architect or quantity surveyor must be certain that the contractor has in fact incurred direct loss and/or expense from the event which he is relying on. In the example, the contractor might well not be able to prove that all his prolongation costs were directly attributable to the strike. It is on this second point that many contractor's claims fail.

'Direct loss and/or expense' and 'expense'

Under GC/Works/1 claims are to be made on an 'expense' basis in contrast to those under JCT where reimbursement is of 'loss and/or expense'. JCT 80, clause 26.1 provides for reimbursement if the contractor 'has incurred or is likely to incur direct loss and/or expense' for which he will not be reimbursed under any other provision in the contract.

'Direct loss and/or expense' means that the sums recoverable are equivalent to damages at common law: *F.G. Minter Ltd* v. *WHTSO* (1980). The conjunction 'and/or' is used by the draftsman to remove any possible doubt as to the scope of the contractor's entitlement. The contractor is given two separate heads of claim, i.e., actual losses incurred as a direct result of the circumstances giving rise to the entitlement *and* disbursements and other expenditure occasioned as a direct result.

Edition 2 of GC/Works/1, clauses 9(2)(a)(i) and 53 provide for claims if the contractor 'properly and directly incurs any *expense* beyond that otherwise provided for in or reasonably contemplated by the contract'. The last test is an objective one; it is not the contemplation of the particular contractor. At one time I was of the view that *expense* under GC/Works/1, Edition 2, was limited to moneys paid out, which made me popular with the PSA if not with contractors, but I have recanted of that heresy, as has the PSA.

The word must be interpreted in the context of the contract, and in law the word 'expense' is not necessarily treated as excluding loss in the wider sense. So, in a different context, the Court of Appeal rejected the argument that the words 'at the expense of the deceased' inferred that there must be something like the putting of his hand in his pocket by the person who extinguished the debt, and upheld the trial judge's view that in their natural and ordinary meaning these words were 'plainly covered by the definitions contained in the *Oxford Dictionary* of "expense"; the words "cost or sacrifice involved in any course of action" and so on are used.': *Re Stratton's Deed of Disclaimer* (1957) *per* Lord Justice Jenkins. On different wording the Court of Appeal took a more restrictive view in the charterparty case of *Chandris* v. *Union of India* (1956) where Lord Justice Denning was of the opinion that a clause in a charterparty entitling the shipowner to recover 'any expense in shifting the cargo' was to be interpreted strictly. He said:

> 'The word "expense" means money spent out of pocket and does not include loss of time. At any rate, that is clearly its meaning in this charterparty because there are numerous clauses which draw a

distinction between "expense" or "expenses" on the one hand and "time occupied" on the other . . .'

The former view applies to GC/Works/1, Edition 2, because the other clauses in the contract do not clearly make a distinction between 'expense' and 'loss'. Clause 9(2)(a)(ii) refers to 'cost' and if this is to be distinguished from expense, it is clear that 'expense' has a wider meaning. Furthermore, in the Property Services Agency's (PSA) *Notice to Tenderers – Valuations under Conditions 9 and 53* (C2041, July 1986), it was recognised that 'expense' includes, amongst other things, 'interest not earned if the contractor uses his own capital', thus giving support to the wider meaning. Expenses actually incurred would in any case include any true additional overhead costs to the contractor, e.g., additional supervision, the cost of keeping men and plant on site; and I am now firmly of the view that the *Oxford Dictionary* meaning cited earlier is to be applied under Edition 2.

Contractors often allege that the PSA's implementation of its directive does not always accord with the general principles of law laid down in the *Minter* case because of its interpretation of clause 9(1). The PSA accepts that where any prolongation or disruption is caused by any matter referred to in clause 53(1) the contractor is entitled to recover 'expense' (including finance charges) under that clause. But it also interprets clause 53(1), when it relates to variation instructions, as being limited to the disruptive effect of variations on other non-varied work.

The PSA approach is that the value of the varied work itself (including disruption and finance charges, if any) is to be valued under clause 9(1). This is because there is necessarily delay between the execution of any work and the contractor being paid for it, and so the contractor's original rates must be taken to include an element for finance charges to cover that delay.

How this principle is applied is a matter of dispute between contractors and the PSA. The principle seems sound, but the length of the period of delay is open to question. It is unreasonable to expect contractors to have included in their rates for financing the varied work over an indeterminate period which he cannot have anticipated since it was caused by a variation order. In other words, the valuation of the variation must be undertaken within a 'reasonable' period, which must be determined by reference to the factual situation.

Happily or unhappily, the situation has been resolved under Edition 3 of GC/Works/1 which has been radically revised. In the first place, the method of dealing with prolongation and disruption caused by variations used by the PSA from July 1986 has been given express contractual sanction since the new rules for the valuation of both variation and other instructions (clauses 41–43) must include for not only the direct cost of complying with the variation instruction (VI) or other instruction, but also the cost of 'any disruption to or prolongation of varied and unvaried work consequential on' compliance using the rates or prices in the bills. Secondly, clause 46 (the equivalent of Edition 2, clause 53) is now confined to such disruption and prolongation caused by such things as failure to supply design information and other defaults which are the authority's responsibility, and it is the contractor's responsibility to submit a detailed claim upon which the quantity surveyor will decide. Finally, Edition 3 introduces a restrictive definition of 'expense' as being 'money expended by the contractor'. It does 'not include any sum expended, or loss incurred, by him by way of interest or finance charges however described': clause 46(6).

Later chapters discuss some of these points in more detail, along with various aspects of the contract machinery. A major issue with all contractual claims is the exact amount of money due. The golden rule is to look at the contract to see what supporting evidence the contractor must supply in support of his claim, what procedures must be followed, and what the duties of the contractor, contract administrator or quantity surveyor are. All too often the 'claims clauses' are looked upon as a means of making more money; that is not their purpose: that is to reimburse the contractor for costs or losses which he has suffered as a direct result of the specified events some, but not all, of which may entitle him to bring an alternative claim at common law.

Chapter 2

Practice and Procedure

Notice requirements

All standard form construction contracts require the giving of notices at various stages as work progresses. In some cases the service of notice, or the making of an application, is a precondition to entitlement under the contract. In other cases, getting something in writing is not mandatory, but is a sensible business precaution. One must always study the wording of the relevant contract clause to determine exactly what is required.

So, for example, there is a contrast between the requirements of JCT 63, clause 23, and JCT 80, clause 25, which deal with extensions of time. Under clause 23 the contractor is not required to give notice of delay which will be caused by some expected future event. However, it seems that he is required to give notice if there will be inevitable delay because of some event which has already happened. This was so held in the now well-known case of *London Borough of Merton* v. *Stanley Hugh Leach Ltd* (1985). The notice must specify any delay which has started to affect progress and must not relate to an anticipated future delay.

The position under clause 25 of JCT 80 is different because the contractor's duty to give notice arises 'if and whenever it becomes reasonably apparent that the progress of the works *is being or is likely to be delayed* . . .' In other words, the contractor must given notice of delays which are anticipated and which are likely to affect progress. In neither case does the contract lay down a special form for the notice. All that is needed is that the document should convey the required information and serve to put the architect on notice.

The *Leach* case is very helpful to contractors because of what was said about clause 23 and 25 notices and other matters. For instance, the employer's contention that the contractor must go into abundant detail was utterly rejected, both by the arbitrator and the judge. Merton had argued that if, for instance, the cause of delay relied on was late information, the contractor must specify in his notice what it is the non-receipt of which is causing delay.

That contention was not accepted. Mr Justice Vinelott agreed with the arbitrator's statement that 'the intention of the contractor's notice is simply to warn the architect of the current situation regarding progress. It is then up to the architect to monitor the position in order to form his opinion'. The judge, however, added that if the contractor fails to give the notice at the right time, 'he is in breach of contract and that breach can be taken into account by the architect in deciding whether he should be given an extension of time'. The contractor's duty is to give notice when it is reasonably apparent to him that progress of the works is delayed (or, under JCT 80, also when it is likely to be delayed). He must give the architect as much information as he can as to the cause of the delay.

Just as the contractor is bound to give certain notices, so under most contracts it is the duty of the architect or engineer to issue certificates and to give notices. These must be signed by somebody, and – in the public sector at least – they will often be signed by a person other than the designated architect or engineer. Often he will be referred to by his corporate job description while in fact administration of the contract will be in the hands of a job architect or another employee of the department concerned.

Contrary to the commonly held view, this practice in no way affects the validity of any certificate or notice. Unless the contract states that a certificate or notice shall be *personally* signed by a stated person, the law does not require that he should personally sign the notice. But it must be signed in his name: *London County Council* v. *Vitamins Ltd* (1955). Although that was a landlord and tenant case, the ruling is of general application.

A signature by procuration ('per pro' or pp) or by proxy is quite valid. Lord Justice Romer put the matter very clearly:

> 'It is established as a general proposition that at common law a person sufficiently "signs" a document if it is signed in his name and with his authority by somebody else, and in such case the agent's signature is treated as being that of his principal.'

In that case, the signature on a notice to quit was not invalidated by the omission of the abbreviation 'pp'. Provided the signature on a document is duly authorised, it will be valid. Oddly enough, in an earlier case, the Court of Appeal suggested that the use of a rubber stamp, while valid, was undesirable as a matter of proper practice: *Goodman* v. *Eban Ltd* (1954).

The result in practice is that unless the word 'personally' or some similar expression is used in the contract, any certificate or notice may be signed by an agent acting with the authority of the named person. It is then irrelevant whether the agent adds words indicating that he is signing as proxy or agent so long as he is specifically authorised to sign.

Commonsense dictates that the architect or engineer should notify the contractor of his authorised representatives. An example of a suitable letter (and many other useful drafts) will be found in David Chappell's *Standard Letters in Architectural Practice* (1987, Architectural Press Ltd).

Even where the contract does not require it, prudence dictates that important notices, etc. should be sent by recorded delivery or registered post, although it has been held that even where the contract requires the use of recorded delivery, failure to observe that requirement is not fatal. Notice can, of course, be served personally on the person to whom it is addressed, but is usually sent by post.

Proof of posting is not proof of delivery, but raises a presumption of delivery which stands until rebutted. This is a question of fact: *Gresham Estate Co.* v. *Rossa Mining Co.* (1870).

One's experience of the industry emphasises that both the design team and contractors need to sharpen up on the administrative side, as employers and contractors alike can lose out if what the contract requires is ignored.

Contemporaneous records

Like most standard form construction contracts, the ICE Conditions for Civil Engineering Works (5th edition) lays down procedures for the contractor to give notices and make claims. Many otherwise valid claims are ignored or disputed because the contractor fails to do what the contract requires. The contractor's site agent or representative is the key man in a claims situation, and all too often other pressures mean that this vital aspect of his job is either neglected or ignored. This can be costly to his employer.

Clause 52(4) is a general notice provision. Apart from variations of work and changes of quantities, the contractor must give written notice of his intention to claim to the engineer 'as soon as reasonably possible after the happening of the events' which give rise to the claim. Although the courts might give a generous interpretation to this phrase, this cannot be relied on, especially as it is then the contractor's duty to keep contemporary records to support his claim.

He must do this from the moment the events occur – and so it is back to the man on site. Once the contractor's notice is given, the ball is in the engineer's court, and he has the right to ask for further records to be kept by the contractor and have copies of them. The contract does not say that the engineer can dictate the form which the records should take, but this seems implicit in the wording. The emphasis is on contemporary records and what is happening on site is the vital factor.

The claims procedure goes on to say that, after giving notive, it is for the contractor to submit to the engineer an interim account giving full and detailed particulars of the amount claimed and of the grounds for the claim. What the contract says is that the interim account must be sent 'as soon as is reasonable in all the circumstances', and in practice this means sooner rather than later. The circumstances to be considered are the engineer's as well as those of the contractor. After the contractor's first interim account, the engineer will tell him how frequently further accounts are needed.

The contractor's failure to do what clause 52(4) requires is not necessarily fatal to his claim. If he fails to give notice or keep and provide copies of the relevant records, the contractor is entitled to payment 'only to the extent that the engineer has not been prevented from or substantially prejudiced by such failure in investigating the claim'. Otherwise he is entitled to prompt payment of the monies due to him in the normal certificates.

Engineers will often exclude part of the claim on the grounds that they have been prevented or prejudiced from investigating the claim. In that case, it is clear that the contractor is entitled to submit further proof of the claim at any time before the final certificate. This follows from the wording of clause 52(4)(e):

> 'If the contractor fails to comply with any of the provisions of this clause . . . then the contractor shall be entitled to payment in respect thereof only to the extent that the engineer has not been prevented from or substantially prejudiced by such failure in investigating the said claim'.

This does not set any time limit on the engineer's investigation, but does mean that the contractor has to resubmit his claim and has lost the benefit of early settlement. This should not be necessary if the contractor has kept 'such contemporary records as may reasonably be necessary to support' his claim and done everything else that clause 52(4) requires.

If the claim is disputed, the contractor's remedy is to refer the matter to arbitration under clause 66, but this is seldom desirable or necessary. It should not be necessary if the claim is a valid one and the contract requirements have been complied with. But suppose that the contractor's claim has been rejected on the ground of prevention or prejudice and the contractor invokes clause 66? In that case, the situation is neatly put by John Uff QC in Keating's *Building Contracts*, 4th edition, p. 509:

> 'Where a claim has been rejected . . . the dispute to be decided under clause 66 is whether the engineer . . . *has been* prevented or prejudiced. Both the arbitrator and the engineer under clause 66 are limited to such dispute, and it is irrelevant whether they are similarly prevented or prejudiced. However, they are bound . . . to consider all information made available to the engineer before the final certificate, whether or not there has been a formal resubmission of the claim under the contract'.

But arbitration is a costly business, even under the ICE procedure, and is the last resort. Many arbitrations about claims could be avoided if the contractor and engineer operated the claims machinery correctly. It is my experience that the majority of engineers do so, but the engineer is quite entitled to reject that part of a contractor's claim which he cannot investigate properly because of lack of detailed records.

The majority of claims clauses in the 5th edition of the ICE Conditions are governed by the provisions of clause 52(4) which in fact says that the contractor must observe its procedure where he 'intends to claim any additional payment pursuant to any clause of these conditions other than' clauses 52(1) and 52(2). It is a general notice provision which triggers off the claims machinery, but all too often what it says is ignored.

The cost of keeping the contemporary records required is that of the contractor, even where the record-keeping results from the engineer's requirements, and contractors (and their site representatives) must accept that, subject to the exceptions mentioned, all claims 'pursuant to' the Conditions are included in the requirement. In effect, only common law claims for breach of contract or other ex-contractual claims are excluded.

Good record-keeping is a major factor in a successful claim and the man on site bears the heaviest responsibility. Major contractors issue procedure manuals to their personnel so as to ensure that there is no dispute about substantiated claims. The engineer has no power to compromise disputed claims; what he must do is to determine what is due to the contractor on the basis of the evidence supplied.

Claims under GC/Works/1: Edition 3

The money claims clause in Edition 3 of GC/Works/1, published in December 1989, is a very limited provision when compared with its predecessor, although the principles behind it remain the same.

Variations, which are the main source of disruption and prolongation claims, are excluded from clause 46 which is restricted to 'breach of contract' type claims. The rules for valuing variations now include for the disruptive effect using the rates and prices in the Bills of Quantities, and valuing the disruption consequent on the issue of a VI upon unvaried work as well - a point which other contract draftsmen might adopt.

As under Edition 2, the contractor's entitlement is to recover 'expense'. This is defined [clause 46(6)] as meaning money paid out by the contractor, but it does not 'include any sum expended, or loss incurred, by him by way of interest or finance charges however described'. Limited provision for payment of finance charges is made in clause 47.

Contractors will certainly need to tighten up their procedures because the provisions are tight. I can say that I do not share the view that the clause is 'harsh' or 'unfair'. But certain points must be understood:

- The expense must be one which is 'beyond that provided for or reasonably contemplated by the contract'.
- It must also have been one 'properly and directly incurred' and one which the contractor 'would not otherwise have incurred by reason of' specified matters.

These matters are:

- Direct works carried out by the employer at the same time as the contract works.
- Delay in giving possession of whole or part of the site.
- Late design information which it is the project manager's responsibility to provide.
- Work or supply of goods which is to be ordered direct by the employer or work or supply of goods to be undertaken by the employer, except where this arises through the contractor's default. This ground is conditioned by clause 46(4) since it is only applicable where the employer has failed to supply an item or do something by a date agreed beforehand with the contractor or within any reasonable period specified in a notice given to the employer or project manager by him.

- The employer's or project manager's instructions 'regarding the nomination or appointment of or admission to the site or issue of any pass to any person or any instruction or consent of the' employer under clause 63(2) [Nomination].

Further limitations are imposed by clause 46(3). It is a condition precedent to any claim that:

- *Immediately* 'upon becoming aware that . . . regular progress . . . has been or is likely to be disrupted or prolonged' the contractor must have given written notice to the project manager.
- This must specify the circumstances causing or expected to cause the disruption or prolongation *and* that he is, or expects to be entitled to an increase in the Contract Sum under clause 46(1).
- Clause 46(1) itself imposes a further restriction, namely that the specified event must result in regular progress being *materially* disrupted or prolonged, thus excluding trivial interruptions.
- As soon as reasonably practicable after incurring the actual expenses, *and in any case within 56 days of incurring them,* the contractor must provide the quantity surveyor with full details of them and evidence that all the expenses directly result from one of the specified events.
- A tight timescale is also imposed on the quantity surveyor [clause 46(5)] who must notify his decision to the contractor within 28 days of receipt of the relevant information from him. These periods can be extended by agreement under clause 1(4). Moreover, alternative claims at common law are not excluded if, for some reason, a contractor misses the boat.

Amounts determined under clause 46 are to be included in the next monthly advance on account: clause 48(2)(d).

Clause 47 is a sensible attempt to deal with the problem of finance charges. It lays down that they are payable only where the employer, the project manager or the quantity surveyor has failed to comply with any time limit specified in the contract and as a result money is withheld from the contractor or where the quantity surveyor varies a decision of his which has been notified to the contractor: clause 47(1). Finance charges are payable as a percentage of the sums which would have been paid to the contractor at a rate of 1% over the rate charged during the relevant period by the Bank of England for lending money to the clearing banks. They are added automatically to the money due.

Finance charges are not payable if they result from:

- Any act, neglect or default of the contractor or any sub-contractor.
- Any failure by the contractor or any sub-contractor to supply the project manager or quantity surveyor with any relevant information.
- Any disagreement about the final account.

The effect of clause 47(6) is to prevent the contractor from claiming interest as 'special damages' for breach of contract under the principles laid down in a series of cases culminating in *President of India* v. *Lips Maritime Corporation* (1987) and *Holbeach Plant Hire Co. Ltd* v. *Anglian Water Authority* (1988).

A case for sharpening up claims correspondence

One of the peculiarities of the ICE Conditions of Contract, 5th edition, is the role of the engineer under the clause 66 disputes procedure. Before arbitration of any dispute, it must 'be referred to and settled by the Engineer who shall state his decision in writing . . .' The perils in operating clause 66(1) are shown by *ECC Quarries Ltd* v. *Merriman Ltd* (1988). Once the disputes process is set in motion by either party, time may run out against him and deprive him of any remedy.

The case arose out of a contract in ICE form made on 29 April 1985 under which Merriman undertook to remove a quantity of overburden from a quarry. On 11 August 1985 it made a claim for extra payment under clause 12 because it had run across hard material which could not reasonably have been foreseen by an experienced contractor. The engineer agreed that extra payment was due and, on 23 December 1985, the engineer agreed Merriman's claim in writing in the sum of £43,000. On 30 January 1986 ECC, the employer, refused to pay the sum agreed, alleging that the hard material was reasonably foreseeable.

Merriman wrote to the engineer on 19 February 1986 referring to the employer's refusal and requesting that:

> 'you give notice to both the employer and ourselves under clause 66(1) . . . as to your decision on the outstanding payment and therefore to either certify payment of the outstanding sum or refuse to do the same'.

Subsequent to a 'without prejudice' meeting in July 1986, the engineer wrote to Merriman on 31 July:

'In connection with your clause 12 claim we have considered again all of the evidence available and have come to the conclusion that it is not correct to arrive at a valuation by way of considering the total number of hours worked by the D9 ripper and taking a percentage, as discussed with you previously'.

He then quantified the sum due as £12,587 and said that he had issued a certificate for that amount, without specifying when or how the certificate had been issued.

Merriman gave notice of arbitration. The employer sought a declaration in the High Court that Merriman was not entitled to refer the claim for extra payment to arbitration on the grounds that it had failed to comply with clause 66(1) and that the letter of 31 July 1986 was the engineer's decision under clause 66(1), even though it did not specifically describe itself as such. Merriman's notice of arbitration was not given within three months of the letter.

His Honour Judge John Davies QC accepted the employer's argument. He held that the letter of 31 July was a decision of the engineer under clause 66(1), and the contractor had no right to arbitration. He said:

'If the contractor had any doubt regarding the true status of the letter he had until 19 August, 1986 to refer the matter to arbitration on the grounds that the engineer had failed to give his decision within three months of being requested to do so. Alternatively, he had three months from 31 July 1986 in which to do so on the assumption that (the letter) was a valid, clause 66(1) decision'.

The whole purpose of clause 66(1) and its procedure is, of course, to ensure that neither employer nor contractor is caught unawares by the strict time limit for arbitration as well as giving the engineer a chance to reconsider matters. The point is well made by Max Abrahamson in *Engineering Law and the ICE Contracts*, 4th edition, p. 292:

'The contractor should make it quite clear when he is applying for an engineer's decision and not merely negotiating, and the engineer should state specifically that he is making a decision under this clause and send a copy of that decision to both parties'.

In this case, both parties may well be forgiven. What is perhaps surprising is that wider-ranging legal arguments were not raised. The judgment refers to the engineer having 'agreed' the original sum due as being £43,000. On this basis it might have been argued that the employer was precluded from objecting to the figure on the basis of estoppel: see *Rees & Kirby Ltd* v. *Swansea City Council* (1983). Interesting questions of the engineer's authority as agent might also have been canvassed.

As it is, *ECC Quarries Ltd* v. *Merriman Ltd* is helpful in considering clause 66(1) situations, and the learned judge's approach may be contrasted with that of the Court of Appeal in *Monmouthshire County Council* v. *Costelloe & Kemple Ltd* (1965) which arose under the ICE Conditions, 4th edition. There, an engineer's letter enclosing his 'observations and comments' on a contractor's claims, all but one of which he rejected with the words: 'I cannot agree with or consider this claim', in reply to a contractor's letter requesting comments, was held *not* to be a decision of the engineer so as to start time running.

Lord Justice Harman made the point that clause 66 is one which can deprive contractors

> 'of their general rights at law and therefore one must construe it with some strictness as having a forfeiting effect. It is not a penal clause, but it must be construed against the person putting it forward who is, after all trying to shut out the ordinary citizen's rights to go to the courts to have his grievances ventilated.'

That view does not represent the current climate of judicial opinion, and what clause 66(1) does is to prevent the dispute being arbitrated once the very short time limit has run. The *Monmouthshire County Council* case still provides valuable guidance into how clause 66(1) should be analysed, while the Merriman decision serves as a warning to contractors and employers alike. On one view, the case discourages negotiated settlements – and certainly contractors must appreciate the importance of initiating arbitration in time.

Merely submitting a claim does not mean that there is a dispute or difference, which requires a definite issue, e.g. outright rejection of a claim, followed by a request for adjudication under clause 66(1). Engineers and contractors must sharpen up their claims correspondence.

JCT 80: clause 26 requirements

Clause 26.1 of JCT 80 sets out the procedure which must be followed if the contractor wishes to obtain reimbursement of direct loss and/or expense which he has incurred or is likely to incur due to deferment of possession of the site or because regular progress of the works has been or is likely to be materially affected by one or more of the events listed in clause 26.2. The use of the word 'materially' in clause 26.1 excludes trivial interruptions. Clause 26.1 does not require the contractor to make an application, but it governs the steps to be taken if notice is given. If written application is made, it is quite clear that the contractor must particularise his entitlement and this duty arises at two stages.

Firstly, the written application must give the architect sufficient information to enable him to form an opinion that the contractor has been or is likely to be involved in qualifying loss and/or expense by reason of the specified events. If the contractor's application contains insufficient detail, the contract confers a right on the architect to seek additional information from the contractor: see clause 26.1.2.

The second stage arises if the architect's opinion is affirmative and further information is then needed to enable the architect (or the quantity surveyor if that duty is delegated to him) to ascertain the amount of loss and/or expense. 'The contractor shall submit to the architect or quantity surveyor upon request such details of such loss and/or expense as are reasonably necessary' for the purposes of the ascertainment: clause 26.1.3.

This requires much more detailed supportive information from the contractor; it will not be available until the loss and/or expense is capable of calculation. Under a contract in JCT 63 terms, the position was put in this way by Mr Justice Vinelott in *London Borough of Merton* v. *Stanley Hugh Leach Ltd* (1985):

> 'If the architect on application by the contractor forms an opinion favourable to the contractor it is his duty to ascertain or to instruct the quantity surveyor to ascertain the loss or expense suffered. The machinery of the contract for payment of the contract sum and in particular the payment on issue of interim certificates then applies. The contractor must clearly cooperate with the architect or quantity surveyor in giving such particulars of the loss or expense claimed as the architect or quantity surveyor may require to enable him to ascertain the extent of that loss or expense; clearly the contractor cannot complain that the architect has failed to ascertain or to instruct the quantity surveyor to ascertain the amount of direct loss or expense attributable to one of the specified heads if he has failed adequately to answer a request for information which the architect requires if he or the quantity surveyor is to carry out that task.'

The position is the same under JCT 80 and, indeed, under IFC 84.

If the contractor invokes clause 26 and does what is required, the architect is under a *duty* to ascertain or instruct the quantity surveyor to ascertain whether loss or expense is being incurred and its amount. This follows from the wording of clause 26.1 which uses the word 'shall' and so imposes a duty on the architect, provided that the architect has formed a prior opinion that the contractor has been or is likely to be involved in direct loss and/or expense as a result of the specified event(s) and which is not recoverable under any other provision of the contract.

There is no doubt that the employer is liable in damages for breach by the architect of this duty and this is so whether the architect is an employee, e.g., where the employer is a public authority, or as is more usual, an independent architect engaged by the employer. Where clause 26.1 says 'the architect shall' this in effect means 'the employer shall procure that the architect shall'. This point is implicit in the reasoning in *Merton* v. *Leach* and also follows from *Croudace Ltd* v. *London Borough of Lambeth* (1986) which is clear authority for the view that the architect's failure to ascertain, or instruct the quantity surveyor to ascertain, the amount of direct loss and/or expense suffered or incurred by the contractor is a breach of contract for which the employer may be liable in damages if the contractor can establish that he has suffered damage as a result of the breach.

This he can do without difficulty in most cases and it does not raise any great difficulties. In *Croudace* Lord Justice Balcombe dealt with the matter in this way:

> 'Unless it can be successfully maintained by Lambeth that there are *no* matters in respect of which Croudace are entitled to claim for loss and expense under [what is now clause 26], it necessarily follows that Croudace must have suffered *some* damage as a result of there being no one to ascertain the amount of their claim . . .'

In that case the employer had failed to appoint a successor architect when the named architect retired.

A general analysis of the contract shows that the employer, although one of the two contracting parties, has little opportunity to take any active part in its administration, and most of the acts and decisions under the contract are specifically required to be taken by the architect. The contract would be unworkable without an architect and it is for this reason that the employer will be liable if the architect fails to perform his mandatory duties, whatever the legal nature of that obligation may be.

GC/Works/1: Edition 3 – extensions of time

It would be remiss of me not to deal with the revised provisions governing the grant of extensions of time in Edition 3 of GC/Works/1, 1989 since claims for time are endemic on any building contract. Under Edition 3, extension of time is covered by clause 36 which has been considerably revised. Procedurally, it is excellent, but contractors will not appreciate the most radical change, which is that there is no entitlement to an extension because of bad weather.

The rationale of this is that the contract periods stipulated in the tender documents have been calculated with an allowance for anticipated weather conditions. In tendering, the contractor has priced on that basis and he must work on the assumption that the actual weather may be better or worse than that allowed for by the Authority.

Clause 36(1) says that where the project manager 'receives notice requesting an extension of time from the contractor or *where he considers that there has been or is likely to be a delay which will prevent completion . . .* by the relevant Date for Completion', he must consider whether any extension of time is merited. In most cases, therefore, the project manager will be considering whether to grant an extension of time after he has received notification from the contractor. But the italicised phrase makes clear that he can grant an extension in the absence of a request from the contractor. The project manager must, indeed, make an award if delay is caused by some act or default for which the Authority is responsible and which is bound to delay completion as otherwise time may become 'at large'.

Nothing is said about the content of the contractor's notice, but I think that it must give details of the circumstances that have or might cause delay and also an estimate of the likely delay. The project manager must notify the contractor of his decision 'as soon as possible and *in any event within 6 weeks from the date of any notice he has received'* from the contractor. This time-limit is strict and cannot be waived. The project manager is bound to say whether his decision is interim or final: see clause 36(3).

The project manager must keep any interim decisions under review and he must reach a final decision on all outstanding and interim extensions of time within 42 days after completion of the works. Completion of the works is a cut-off point since no requests for extensions can be submitted by the contractor thereafter. In his final decision, the project manager is not entitled to withdraw or reduce any interim extensions already awarded.

If the contractor is dissatisfied with the project manager's decision under clause 36(1), he has 14 days from its receipt in which to submit a claim to the project manager specifying the grounds which in his view entitle him to an extension or further extension of time. The project manager has 28 days from receipt of the contractor's claim in which to reconsider the matter and notify the contractor. Once again all these time periods are strict.

Although clause 36 is silent on the matter, the intention plainly is that the project manager should give reasoned decisions and I think that his award should refer to the appropriate sub-paragraphs.

Contractors and project managers alike must remember that the grant of extensions is not an exact science, though some claims consultants take the opposite view. Whereas a money claim can be ascertained precisely, a claim to time cannot, and the project manager must do the best he can on the information available to him.

Clause 36(7) is important and has its parallel in JCT 80. The contractor loses his entitlement to an extension of time where 'the delay or likely delay is, or would be attributable to, [his] negligence, default, improper conduct or lack of endeavour' (clause 36(7)). Under the same sub-clause, the contractor is also bound to 'endeavour to prevent delays and minimise unavoidable delays, and to do all that may be required to proceed with the works' – an obligation which is more clearly expressed than its JCT 80 equivalent and which plainly does not contemplate the expenditure of substantial sums of money. He must not sit back passively but is required to act reasonably.

Clause 36(2) provides that the project manager can only award an extension if he is satisfied that the actual or likely delay is due to specified events. They are:

- The execution of any modified or additional work.
- The act, neglect or default of the employer or the project manager.
- Any strike or industrial action which prevents or delays the execution of the works and which is outside the control of the contractor or any of his sub-contractors (including nominated sub-contractors).
- An Accepted Risk, i.e., insurance risks which are given a limited definition in clause 1(1).
- Unforeseeable ground conditions notified to the project manager under clause 7(3). They are defined there as 'ground conditions (excluding those caused by weather but including artificial obstructions) which [the contractor] did not know of, and which he could not reasonably have foreseen having regard to any information which he had or ought reasonably to have ascertained'.
- Any other circumstances (other than weather conditions) which are

outside the control of the contractor or any of his sub-contractors and which could not have been reasonably contemplated under the contract. This broad 'sweeper' must be given a restrictive interpretation; it cannot be used merely as a convenience and it does *not* cover delays caused by nominated sub-contractors.

ICE Conditions: the importance of notice

The standard form construction contracts require the contractor – usually when he wishes to make a claim for extra time or money – to give notices to the architect or engineer at the proper time.

Recent case law suggests that the courts will adopt a sensible and business-like approach to the matter, but this should not lead contractors into thinking that the giving of notices is a mere formality or something which can be overlooked.

The point is neatly illustrated by *Humber Oils Trustees Ltd* v. *Hersent Offshore Ltd* (1981), a case arising out of a contract which incorporated the ICE Conditions (4th edition). The work involved retrieving old steel piles from the sea-bed, welding on new sections and redriving the piles.

Clause 12 of the ICE Conditions, 4th edition, said that if the contractor intended to claim extra payment for unforeseeable physical conditions or artificial obstructions, he was to give notice to the engineer

> 'specifying (a) the physical conditions and artificial obstructions encountered and (b) the additional work . . . and constructional plant which he proposes to do and use . . . and (c) the extent of the anticipated delay or interference with the execution of the work.'

There then followed an important proviso:

> *'Provided always that the cost of all work done or constructional plant used by the contractor prior to the giving of such notice . . . shall be deemed to have been covered in the rates and prices referred to . . .'*

A further provision [clause 12(3)] said that if, at the time of giving his notice,

> 'the additional constructional plant or work which the contractor proposes to use or do are then sufficiently defined to enable the contractor to give a quotation for the payment of the cost thereof . . .'

he was to prepare and submit a quotation with his notice and

> 'in all other cases . . . (to) submit with such notice an estimate of the additional constructional plant and an estimate of the cost of the . . . delay or interference.'

As work progressed, the welds were subject to cracking, thought to be due to the unexpected presence of hydrogen in the old piles. This caused considerable delay and extra expense and the contractor sent the engineer a notice under clause 12.

This merely referred to 'the extra costs involved in the solving of the welding problem and the subsequent extra works required to provide butt welds'. It gave no quotation or estimate as required by clause 12(2), but added that costs would be monitored 'and a full report and claim, detailing the extra plant and cost, will be submitted in due course'.

The contractor subsequently gave a further notice specifying what it intended to do, but stating only that 'the claim will contain the cost of all these items . . . and also the cost of all the major items that are affected by the delay in the programme'.

When the contractor later submitted its claim, the engineer rejected it on the basis that the appropriate notice failed to satisfy the requirements of clause 12(2). The court upheld that rejection and noted that the requirement for quotations or estimates, which had also not been complied with, was also a precondition to the contractor's right to claim. Because no notice had been given as required by the clause, the whole of the additional costs were deemed to have been covered in the contract rates or prices.

This is probably an extreme case – and of course the wording of ICE clause 12 has now been changed – and later cases (on other contract forms) suggest that the courts may well adopt a less legalistic approach.

In *Humber Oils* the judge rejected an argument based on the so-called 'philosophy of the clause', but the heart of the case was the proviso to the clause which is quoted above. Had it not been for that proviso, the claim itself might well have been considered sufficient notice.

The lesson for contractors is to do what any claims clause specifies, as often the timing (as well as the content) of the notice is vital. This applies to claims for both money and extension of time, whether under ICE Conditions or general building contract forms such as JCT 80.

The case of *London Borough of Merton* v. *Stanley Hugh Leach Ltd* (1985) is important on the timing and content of notices under JCT contracts where, under JCT 63 at least, a more generous view may be taken.

Mr Justice Vinelott ruled, among other things:

- A clause 23 notice (extension of time) could be a proper notice even if it did not specify a cause of delay with sufficient detail to enable the architect to form an opinion as to whether the event specified fell within the clause, because the intention of the notice is simply to warn the architect of the current situation.
- But the contractor must give the architect as much information as he can about the cause of the delay, and he must do this promptly.
- Applications for direct loss and/or expense under clauses 11(6) and 24(1) must be framed in sufficient detail. As the headnote in *Building Law Reports* puts it:

> 'The application must be made within a reasonable time; it must not be made so late that the architect can no longer form a competent opinion of the matters on which he is required to satisfy himself that the contractor has suffered the loss or expense claimed.'

JCT 80, of course, lays down a much more detailed procedural code, and a contractor who loses out because his notice is not sufficiently detailed or made at the right time has only himself to blame.

More detail, rather than less, is what is required, and it is quite clear that if the contract conditions require the contractor to take specified steps in a claims situation, neither an arbitrator nor the courts can waive the requirement.

The 'without prejudice' rule

Public policy encourages disputing parties to settle their differences rather than litigate them to a finish. This is the reason for the 'without prejudice' rule which excludes all negotiations genuinely aimed at reaching a settlement from being given in evidence in any subsequent arbitration or litigation. The House of Lords has ruled that the privilege is not lost if a settlement is reached, with the effect that disputing third parties are not entitled to see the settlement correspondence.

This is the key point in *Rush & Tompkins Ltd* v. *Greater London Council* (1988) which arose out of a dispute on a building contract. Rush & Tompkins was the GLC's main contractor, and Careys were ground works sub-contractor, which submitted claims for loss and expense to Rush & Tompkins. The GLC did not agree Careys' claims. Rush &

Tompkins started proceedings against both the GLC and Careys in respect of Careys' entitlement and claiming reimbursement from the GLC. There was a settlement between Rush & Tompkins and the GLC, and while its terms were disclosed to Careys they did not show what valuation had been put on the claim.

Careys believed that the negotiating correspondence must have shown the value of its claims for purposes of the settlement. They applied for discovery of the 'without prejudice' correspondence, i.e. its disclosure, but the Official Referee refused it, accepting Rush & Tompkins' argument that the documents were protected by the 'without prejudice' rule because they came into existence for the purpose of settling the claim. The Court of Appeal reversed his decision and ordered discovery, holding that the privilege ceased once a settlement had been reached.

The House of Lords overturned this ruling and restated the law. Lord Griffiths emphasised that the public policy which protects genuine negotiations from being admissible in evidence must be extended to protect those negotiations from being disclosed to third parties. This is of great importance in a contract claims situation where the employer and main contractor reach a compromise which is unacceptable to a sub-contractor.

Lord Griffiths put the issue succinctly:

> 'The wiser course is to protect "without prejudice" communications between parties to litigation from production to other parties in the same litigation. In multiparty litigation it is a not-infrequent experience that one party takes up an unreasonably intransigent attitude that makes it extremely difficult to settle with him. In such circumstances it would place a serious fetter on negotiations between other parties if they knew that everything would be revealed to the one obdurate litigant. What would in fact happen would be that nothing would be put on paper but this is in itself a recipe for disaster in difficult negotiations, which are far better spelt out with precision in writing.'

Claims negotiations can take place and rely on the 'without prejudice' privilege. It is best to head all negotiating correspondence 'without prejudice' to make clear that in the event of the negotiations being unsuccessful they cannot be referred to in any subsequent proceedings. In fact, the privilege does not depend on the use of the phrase. If it is clear from the surrounding circumstances that the parties were seeking to compromise a dispute, the negotiations cannot be used subsequently to establish an admission or partial admission of liability.

However, the privilege is not absolute. In certain limited circumstances 'without prejudice' material can be looked at if the justice of the case so requires. The House of Lords noted some of these exceptional cases:

- Without prejudice material can be looked at if the issue is whether or not the negotiations resulted in an agreed settlement: *Tomlin* v. *Standard Telephone & Cables Ltd* (1969).
- The court will not allow the phrase to be used to exclude an act of bankruptcy nor to suppress a threat if an offer is not accepted: *Ex parte Holt* (1892); *Kitcart* v. *Sharp* (1882).
- Without prejudice correspondence can be looked at to determine the question of costs after judgment or award: *Cutts* v. *Head* (1904).
- The admission of an independent fact in no way connected with the merits of the dispute is admissible even if the admission was made in the course of negotiations for a settlement, e.g., an admission that a document is in the handwriting of one of the parties.

These are, however, all exceptional cases. They cannot be used to whittle down the protection given to negotiating parties. The vital point is, though, that 'without prejudice' statements and discussions will only be privileged if there is a dispute and an attempt to settle or compromise it. One should not head letters 'without prejudice' indisciminately on the mistaken assumption that this gives one an opportunity to write anything with impunity. It does not.

But where there is a dispute, e.g., as to a contractor's entitlement to extension of time and/or money, and negotiations are in progress, the privilege should be involved. Any admissions made cannot then be used in evidence in any subsequent proceedings connected with the same subject-matter unless both parties consent. Admissions made to reach settlement with a different party in the same proceedings are also inadmissible whether or not a settlement is reached with that party.

The House of Lords' decision was warmly welcomed by the legal profession and by all those engaged in disputes settlement. The fact is that whatever the standard form contracts may provide about payment of claims for disruption or delay, the typical scenario on a long contract is for the global claim to be settled by negotiations at the end of the day, especially where some of the delay is the contractor's responsibility. Common sense and the law go hand-in-hand in protecting the negotiations, even if they prove abortive.

ICE disputes procedure

Some short but interesting points relevant to disputes under the ICE Conditions, 5th edition, were decided by His Honour Judge Lewis Hawser QC in *Wigan Metropolitan Borough Council* v. *Sharkey Brothers Ltd* (1989) which involved a challenge to the jurisdiction of the arbitrator.

Sharkey, the contractor under an ICE contract, submitted an interim account to the engineer in March 1985. It covered measured work and claims. Failing settlement, and after a meeting to discuss the account, Sharkey wrote to the engineer on 19 July 1985 requesting 'your formal decision on the whole of the four outstanding claims and measured items which are currently in dispute'. The engineer asked what items were 'in dispute' on 13 August and the next day Sharkey submitted an amended account identifying the items. The engineer gave his decision on 24 October, and five days later Sharkey replied alleging that some items were still in dispute. Failing satisfaction, they served notice of arbitration on 7 November and listed six matters in dispute, including 'delay, reimbursement of costs' and what was described as a sweeping-up item – 'other matters' – the amount at issue being about £100,000.

Following a meeting before the arbitrator on 1 May 1986, Sharkey sent a letter to the plaintiffs on 19 May in which they identified two other matters which had not been submitted to the engineer for his decision, and in their eventual points of claim they also included eight other items which had not been submitted to the engineer. Sharkey contended that their original reference to 'other matters' was sufficient for the purposes of clause 66 of the ICE Conditions. The arbitrator decided that he had jurisdiction to deal with eight of the unnotified items but ruled that the reference to 'delay, reimbursement of costs' could not include disruption costs.

Wigan's main contention before the court was that there is no room for a so-called 'sweeping-up' clause and that in order to comply with the provisions of clause 66 to give particulars of the matters being referred to the arbitrator in the notice to refer within three months of the engineer's decision. Not surprisingly, Judge Hawser upheld this contention since case law establishes that the clause 66 time limits must be strictly observed unless they are waived or extended by agreement. The Council were entitled to have their attention drawn in clear terms to the matters which were to form the subject-matter of the claim in arbitration within three months from the date of the engineer's decision. The reference to 'other matters' was insufficient and Judge Hawser adopted the views expressed in Mustill & Boyd's *Commercial Arbitration*, pp. 171–2, that a notice to refer 'should give the recipient sufficient information as to the claim which he has to face'. The inadequacy of the reference to 'other matters' was patent, and the position was put beyond doubt by the fact that when requested to elucidate the 'other matters' Sharkey's letter of 19 May 1986 had only mentioned two other matters. Judge Hawser also noted that rule 1(2) of the ICE Arbitration Procedure confirmed his interpretation of clause 66, and most certainly the judgment is in line with the other cases on the provision.

The other interesting point is the judge's ruling on the reference to 'delay, reimbursement of costs' which was that the arbitrator was incorrect since the phrase was 'sufficiently wide to cover disruption' as well as delay. It is not clear what arguments were advanced to support this contention because many arbitrators would have agreed with the Sharkey arbitrator that this was not the case. The ICE Conditions do not provide for the recovery of disruption or prolongation costs as such. The cost of prolongation or disruption is recoverable as part of 'cost' and the headings are or should be distinct; thus, the reference to 'delay, reimbursement of costs' would on general principle seem best limited to prolongation claims. Certainly, it cannot be assumed that the reference will be held to be sufficient in all cases.

Wigan MBC v. *Sharkey Bros Ltd* should now be read alongside such cases as *Monmouth CC* v. *Costelloe & Kemple Ltd* (1965) when considering the operation of clause 66. These cases establish a number of fundamental propositions:

- There must first be a 'dispute or difference', and thus there must be a clear rejection of the claim in unequivocal words.
- Once there is a difference or dispute, it must be referred to the engineer who must give a written decision and notify both employer and contractor.
- The timetable in clause 66(3) is strict. If the time-limits are not met then there can be no valid reference to arbitration. Once the time-limit has expired it is too late to refer further or omitted disputes.
- If the engineer gives his decision within the prescribed period, the dissatisfied party has three calendar months from receipt of it in which to refer the matter to arbitration.
- The notice to refer to arbitration must be explicit and must refer in clear terms to the matters which are to form the subject-matter of the dispute. The respondent is entitled to know what is being alleged and vague and generalised references will be held to be insufficient.

Money claims under ICE Minor Works

The ICE Minor Works Conditions is a good standard form, especially in its treatment of money claims or, as the contract calls them, 'additional payments'. These claims are based on *cost* and the word is given a restrictive definition in clause 3.1. 'Cost' includes 'overhead costs whether on or off the site ... *but not profit*'. The equivalent definition in the ICE 5th edition does not contain the italicised words and they have simply been added to clarify the position. Even under the 5th edition, the legal definition of 'cost' means that profit is excluded.

The Minor Works definition does not apply where the contract is on a cost-plus fee basis and, in the important case of claims arising from adverse physical conditions and artificial obstructions, the contractor gets cost and a reasonable percentage addition for profit.

The general claims provision is clause 6.1. It covers not only additional works but additional cost *'including any cost arising from delay or disruption to progress'* as a result of specified matters. These are:

- Variations, suspension of the works, and changes in their intended sequence ordered by the engineer.
- Testing or investigation where the results are in the contractor's favour.
- Late information from the engineer.
- Employer's failure to give adequate access to the works or possession of land required to perform them.
- Employer's delay in providing the contractor with any materials which he has undertaken to supply.

The procedure is refreshingly simple since, if the contractor carries out additional works or incurs additional cost, 'the engineer shall certify and the employer shall pay ... such additional sum as the engineer *after consultation with the contractor* considers fair and reasonable'. In determining the figure for additional work, the engineer 'shall have regard to the prices contained in the contract', such as in the bills. The value of any omitted work is to be determined by the engineer as well.

There are no complicated notice procedures or restrictive conditions, but it is clear that the contractor must substantiate his claim to additional cost and no doubt the reference to 'consultation' will promote the negotiated settlements which are a happy feature of civil engineering claims generally. But legally and contractually it is for the contractor to prove his claim by means of acceptable evidence.

All the grounds of claim covered by clause 6.1 have in common the entitlement of the contractor to cost only, including overheads. In a prolongation situation, the loss of the right to recover the profit which he might otherwise have earned could be a serious obstacle in some circumstances, but the position is the same under the ICE Conditions 5th edition.

The new Conditions do not exclude the contractor's right to claim damages at common law if the act that is relied on also amounts to a breach of contract, as would the employer's default in supplying materials which he had agreed to supply. Similarly, a damages claim for breach of an implied term could also arise since the Conditions are not exhaustive, and a 'loss of profit' claim could be pursued at common law.

One of the most prolific sources of claims – adverse physical conditions – is dealt with separately by clause 3.8. This is a more straightforward version of its parent in the ICE 5th edition. In this case, payment is dependent upon the contractor serving a written notice on the engineer 'as early as practicable' after he encounters 'any artificial obstruction or physical condition (other than weather)' during the execution of the works.

The contractor is also under the burden of establishing that the obstruction or condition 'could not in his opinion reasonably have been foreseen by an experienced contractor'. If the engineer is of the same opinion, he is to certify 'a fair and reasonable sum' to cover the cost of any necessary additional work, plant or equipment, plus profit, as a result of the contractor's compliance with any instructions he may issue *and/or* his taking proper and reasonable measures to overcome the problem in the absence of instructions from the engineer.

As well as this, the contractor is entitled to 'such sum as shall be agreed as the additional *cost* . . . of the (resultant) delay or disruption'. No profit is allowable on this element of the claim. Failing agreement, it is for the engineer to determine the amount to be paid and it is to be a 'fair and reasonable sum'.

Clearly, the wording of the clause does not restrict payment to costs incurred after the giving of notice by the contractor. The important point is that while the contractor is entitled to payment in respect of 'reasonable profit' on additional work, he is not so entitled in respect of any claim he has for delay and disruption of working. This ties in with the philosophy of clause 6.1 and the contract as a whole.

The contract is silent as to what is 'a reasonable percentage addition in respect of profit', which is also the situation under clause 12(3) of the ICE 5th edition Conditions. Presumably the contractor's own profit levels on the works are to be used as a baseline or at least as a guide, always assuming that those levels are 'reasonable'. Encountering an obstruction or condition covered by clause 3.8 also gives rise to a claim for extension of time under clause 4.4(c).

Clause 3.8 is sensibly less detailed than the equivalent provision in the parent contract, but the engineer's power to give instructions to deal with the problem is widely cast. The clause seems very practical in its approach.

Practicability and simplicity seem to be the keynotes of the Minor Works Condition claims provisions, and neither contractors nor engineers should have any difficulty operating them. They confer valuable benefits upon contractors.

JCT claims machinery

Valuable guidance on the claims machinery of JCT 63 was given by the judgment of Mr Justice Vinelott in *London Borough of Merton* v. *Stanley Hugh Leach Ltd* (1985), and much of what was decided applies to JCT 80 and other contracts as well.

One of the points at issue was whether the contractors had made proper written application to the architect for reimbursement of direct loss and/or expense under clauses 11(6) and 24(1) of the contract. Apparently the contractors applied for reimbursement under both clauses in a single letter or request which did not give much, if any, detail. The employers argued that a contractor's application under either clause must provide the architect with sufficient information to enable him to form an opinion on three separate matters:

- That the regular progress of the works had been materially affected by a specified event under clause 24(1).
- That the contractor had suffered loss or expense thereby or by carrying out a variation instruction.
- That the loss or expense would not be reimbursed by payment under another clause.

The arbitrator had found against Merton on this point for, while he accepted that the architect's contact with the site is not on a day-to-day basis, there are many occasions when one of the specified events occurs which are sufficiently within the architect's knowledge to enable him to form an opinion that the contractor has been involved in loss and/or expense. In his view, although it was desirable for the contractor to give the fullest possible information as part of his application, the contract did not require the contractor to do so. This, of course, is in contrast with the position under JCT 80.

If the contractor's application is confined to the bare minimum requirements of the contract and the architect has insufficient information to enable him to form an opinion, then the architect can and will seek further information from the contractor. 'There is no reason,' stated the arbitrator, 'why this additional information should not be given orally.'

Merton's counter-argument, in summary, was that since the contractor must have grounds for making an application under clauses 11(6) or 24, it was reasonable to interpret those clauses as requiring him at least to make those reasons explicit in his application.

The judge agreed with the arbitrator, and noted that the contractor must act reasonably.

> 'But in considering whether the contractor has [so] acted . . . and with reasonable expedition it must be borne in mind that the architect is not a stranger to the work, and may in some cases have detailed knowledge of the progress of the work and the contractor's planning,' he ruled.
>
> In some cases he pointed out:
>
> 'the briefest and most uninformative notification of a claim would suffice: a case, for instance, where the architect was well aware of the contractor's plans and of a delay in progress caused by a requirement that works be opened up for inspection, but where a dispute whether the contractor had suffered loss or expense in consequence of the delay had already emerged. In such a case the contractor might give a purely formal notice solely in order to ensure that the issue would in due course be determined by an arbitrator . . .'

Despite this ruling, contractors are advised to give the fullest possible information. If they fail to do so, they must be prepared to provide the architect (or quantity surveyor) with information on request. But there is no JCT 63 requirement for 'moneyed-out' claims.

A further blow to architect and employers was the judge's specific ruling that under the money claims provisions, rolled-up claims are permissible. Merton had argued that under clauses 11(6) and 24(1) the architect cannot make an award unless he is in a position to ascertain the direct loss stemming from a specific cause identified in the application, and cannot, therefore, make an award if the loss stemming from two different causes cannot be separated.

Mr Justice Vinelott preferred to follow the decision of Mr Justice

Donaldson in *Crosby* v. *Portland Urban District Council* (1967), whose reasoning he found compelling.

The judge's summary of the JCT 63 position is worth quoting at length:

> 'If application is made for reimbursement of direct loss and/or expense attributable to more than one head of claim and at the time when the loss or expense comes to be ascertained it is impracticable to disentangle or disintegrate the part directly attributable to each head of claim, then, provided that the contractor has not unreasonably delayed in making the claim and so has himself created the difficulty, the architect must ascertain the global loss directly attributable to the two causes, disregarding, as in *Crosby*, any loss or expense which would have been recoverable if the claim had been made under one head in isolation, and which would not have been recoverable under the other head taken in isolation. To this extent the law supplements the contractual machinery which no longer works in the way it was intended to work, so as to ensure that the contractor is not unfairly deprived of the benefit which the parties clearly intend he should have.'

Contractors, of course, can do much to avoid these sort of problems by operating the contract machinery correctly; the 'global approach' is still a last resort. If a contractor decides to make a claim, it is in his own interests to give adequate information. The architect can come back for information to enable him properly to ascertain the claim. As the judge said:

> '. . . if he makes a claim but fails to do so with sufficient particularity to enable the architect to perform his duty, or if he fails to answer a reasonable request for further information, he may lose any right to recover loss or expense . . . and may not be in a position to complain that the architect was in breach of his duty'.

It was also emphasised by the judge that the contractor may have parallel rights to claim compensation under the terms of the contract and to pursue a claim for damages. But that is another story.

Chapter 3

Variations et al

Variations: the need for definition

All building and civil engineering contracts contain a variations clause empowering the architect or engineer to order varied works. Often the circumstances in which a variation can be ordered are widely defined, but plainly a 'variation' must bear some relationship to the work of which it is a variation. Clearly, the architect or engineer cannot order work which is wholly different from the original contract work under the guise of a 'variation'. In practice, though, it is often difficult to decide whether a departure from what the parties expected to be done under the contract is a 'variation'.

This was the point at issue in the case of *Blue Circle Industries PLC* v. *Holland Dredging Company (UK) Ltd* (1987), a Court of Appeal decision on contract conditions which were substantially the ICE Conditions for Civil Engineering Works, 5th edition.

The contract was for dredging work in Lough Larne, Ireland. A special condition (clause 72) provided that 'the coordinated positions of areas within Lough Larne to be allocated for the deposit of the dredged material will be submitted upon approval by the local authorities . . .' Holland's tender allowed for the deposit of dredged material within Lough Larne, and the tender was accepted. 'Had the matter ended there it would have been a perfectly simple offer and acceptance for a dredging contract', Lord Justice Purchas said in the Court of Appeal.

However, while all this was going on, other negotiations were taking place between the parties and various local and statutory authorities, which objected to the discharge of the excavated material from the channel generally into suitable areas of the lough.

The final and agreed solution was that the dredged material should be used to create in artificial kidney-shaped island to be used as a bird sanctuary. The island was of substantial dimensions and involved the construction of a carefully designed bund wall and ancillary works.

On 28 September 1978, Holland submitted a quotation for the 'erection of an artificial island'. On 4 October, Blue Circle accepted the offer by a letter which said that 'an official works order will follow in due course'.

The works order came on 19 October, and there seemed to be two separate agreements in identical terms. The main issue before the court was the relationship between the two agreements and whether the 'island agreement' was a variation of the dredging agreement.

As it turned out, the construction of the island was only a partial success. Holland contended that their offer of 28 September was merely a confirmation of an agreed variation of the dredging contract within clause 51 of the ICE Conditions which, of course, contains no definition of the word 'variation'. Blue Circle argued that the agreement for the construction of the island was quite separate and in its manner of creation was entirely inconsistent with it being a variation under clause 51. The ICE Conditions do not define 'variation'.

Halsbury's *Laws of England*, 4th edition, vol. 4, para 1178, deals with the matter thus:

> 'If the nature or extent of the variation or additional work is such that it is not contemplated by the contract, the contractor can refuse to carry it out or can recover payment for it without complying with the requirements of the variation clause. For varied work to fall outside the contract it must, it seems, either result in it being impossible to trace the original work contracted for or be of a kind totally different from that originally contemplated.'

Earlier case law is to the same effect: *British Movietone News Ltd* v. *London and District Cinemas Ltd* (1952). Lord Justice Purchas preferred the employer's approach. The question to be asked was: 'Could the employer have ordered the work required by it against the wishes of the contractor as a variation under clause 51?'

If the answer to that question was no, then the agreement could not be a variation, but must be a separate agreement. His Lordship put the matter succinctly:

> 'The original dredging contract provided that the spoil . . . should be deposited in Lough Erne "upon approval by the local authorities". In the event, this term of the contract was impossible to fulfil legally. The only alternatives were dumping at sea or the creation of an artificial island. Either of these two solutions was wholly outside the scope of the original dredging contract. Had Holland not been willing, they could not have been obliged to accept the work as a variation.'

In other words, on the facts, there was a separate agreement formed by the exchange of the contractor's quotation and Blue Circle's order. The reference to an order 'to follow' did not make it a conditional contract. The contract was in fact completed, so the court found, when the verbal instructions were confirmed in writing. This is an important decision for users of ICE Conditions and, as Abrahamson, *Engineering Law and the ICE Contracts*, 4th edition, p. 168 has pointed out: 'Most of the employment given to the legal profession by engineering work is to do with disputes about variations'.

A sensible amendment to the ICE Conditions would be to provide a definition of variation. This has been done in the 4th edition of the FIDIC International Conditions (1987) which defines variations as:

(a) increases or decreases of the quantity of work included in the contract;
(b) Omission of such work;
(c) Changes in the character, quality or kind of the work;
(d) Changes in the levels, lines, positions and dimensions of any part of the works;
(e) The execution of additional work of any kind necessary for the completion of the works;
(f) Changes in any specified sequence of timing of construction of any part of the works.

Beware of variations

Variations are a fact of life in the construction industry and give rise to many problems, whichever form of contract is used. All the standard form contracts contain variation clauses because, without an express power to direct departure from the original contract work, the employer would have no power to order it.

To simplify matters, most of the contracts incorporate one or other of the Standard Methods of Measurement, for example JCT 80, clause 2.2.2.1, which now incorporates SMM 7. The supporting provision that 'any departure from the method of preparation . . . or any error in description or in quantity or omission of items . . . shall be corrected' and treated as a variation provides contractors with many opportunities to claim extra remuneration because of the inevitable ambiguities.

The classic case is *C. Bryant & Son Ltd* v. *Birmingham Hospital Saturday Fund* (1938) where the contractor agreed to erect a convalescent home under a contract in RIBA form, 1931 edition. This contained a clause equivalent to JCT 80, clause 2.2.2.1, so that the bills were deemed to have been prepared in accordance with the then current SMM. This required that, where practicable, the nature of the soil should be described, attention drawn to any trial holes, and that excavation in rock should be given separately. The bills referred the contractor to the drawings, block plan and the site to satisfy himself as to local conditions and the full extent and nature of the operations.

Although the existence of rock on the site was known to the architect, it was not shown on any of the plans or referred to in the bills, which contained no separate item for excavation in rock. Mr Justice Lewis ruled that, in the circumstances, the contractor was entitled to treat the excavation in rock as an extra and to be paid the extra cost of the excavation plus a fair profit. Presumably – though this is not clear from the judgment – the rock was not discoverable by an inspection of the site.

This case has been criticised by the present editor of Hudson's *Building Contracts*, 10th edition pp. 518–519, but, with respect, a like result would have been reached had the contract been in JCT 80 form. However, as Hudson points out:

> 'while there remains any doubt as to the possibility of financial claims of this kind . . . the importance of professional preparation of bills in meticulous compliance with the [SMM] is obviously essential in the employer's interest in order to protect him from such claims.'

A quantity surveyor who failed to do this might well find himself on the receiving end of a writ for negligence.

Another case which shows the indulgent attitude of the courts is *A.E. Farr Ltd* v. *Ministry of Transport* (1965), a decision of the House of Lords arising under the ICE Conditions. These provided that quantities were estimated only, that the work was to be remeasured and that unless otherwise shown the bills were deemed to have been prepared in accordance with the CESMM. The contract also said that the contractor was deemed to have satisfied himself as to the correctness of the rates and prices which should 'cover all his obligations under the contract.'

Clause 16 of the bills, which paralleled a clause in the CESMM, dealt with the measurement of excavation in pit or trench and went on:

> 'any additional excavation which may be required for working space . . . will be paid for under separate items, the measurement being the sum of the areas of the sides of the excavation.'

The engineer approved a programme showing the intention to include additional working space in the main part of the job. Clause 16 contained no priced items and there was no item in the bills covering excavation for working space. The House of Lords held – by a majority of three to two – that clause 16, on its true interpretation, was a promise by the employer to pay for excavation of working space because it plainly contemplated work additional to that priced in the bills.

Hudson suggests that this decision is also incorrect, but the true position appears to be as stated in Emden's *Building Contracts*, 8th edition, volume 1, p. 143:

> 'The law would seem to be such that a contractor pricing bills which are to be prepared according to one or other of the standard methods runs almost no commercial risk apart from under-pricing'.

The contractor's errors in pricing are, of course, his responsibility. If the contractor prices too low, for whatever reason, he is entitled only to the bill rate even when those rates are applied to additional work in variations. Bad rates will apply to the varied item so far as is relevant: see P. Hibberd, *Variations in construction contracts* (Collins, 1986), p. 150.

This harsh view is not shared by all the commentators, but seems to be supported by the Court of Appeal decision in *Dudley Corporation* v. *Parsons & Morrin Ltd* (1959), where the contract was in RIBA 1939 form. The contractor priced an item for excavating 575m^3 in rock at 2s (10p) per cubic metre when a reasonable price would have been £2. In the event, 1700m^3 of rock was excavated. The architect valued the work as 2s a cube for 575m^3 and the balance at £2 a cube. The Court of Appeal held that he was wrong to do so, and the contractor was only entitled to 2s a cube for the whole quantity excavated. Even though there was no error by the contractor in pricing in this case, it is clear that the contractor is bound by errors in his pricing (*Higgins* v. *Northampton Corporation* (1927)) both for the original and varied work.

Despite the case law, contractors will no doubt continue to press the view that an erroneous rate can only be applied to the quantity originally envisaged in the contract – and it is certainly not unknown for an erroneous rate to be ignored.

JCT and ICE type variations clauses will continue to give rise to argument, debate and litigation. When the ACA or ACA/BPF or GC/Works/1, Edition 3, forms of contract are used, there is the possibility or prior agreement about the *price* of the variation through the submission of estimates by the contractor, which both contractor and architect must do their best to agree; see clause 17 of the ACA

Contract and the much more precise provisions of GC/Works/1, Edition 3, clause 42(2). If agreement is reached, there will be no quibble over the price (including disruption) or any associated extension of time. But the other problems thrown up by variations remain. No doubt in an ideal world there would be no need for variations and proper pre-planning can still do away with the majority of them. Greater clarity and precision in the drafting of bills would also help.

A one-off situation?

Judicial decisions on the interpretation of a clause in one standard form of contract are not necessarily conclusive when considering the similarly-worded provisions of another standard form. This is the message which emerged from the decision of the Judicial Committee of the Privy Council in *Mitsui Construction Co. Ltd* v. *Attorney-General for Hong Kong* (1986).

The case involved the interpretation of some of the provisions of the former Hong Kong Public Works Department form, which is akin in some respects to the ICE Conditions, 4th edition, as well as FIDIC.

It was concerned with changes in quantity which did not result from variation orders, and the contractor's right to compensation for losses he had sustained because of the extra time he needed to cope with ground conditions.

The contract in question was, in their Lordships' view, 'badly drafted' – the same might be said of most UK standard forms – and the approach of the Privy Council was to interpret the contract on a basis which would attribute 'to the parties an intention to make provision for contingencies inherent in the work contracted for on a sensible and businesslike basis'.

This approach – as the editors of *Building Law Reports* pointed out – is

> 'not entirely objective particularly in a standard form of contract which is a compromise between two conflicting commercial interests and therefore is not always capable of bearing a single "businesslike" meaning'.

The point at issue was whether increases in quantities entitled the contractors to new rates and prices. The works included the excavation and construction of tunnels and the specification laid down that the engineer was to order which type of permanent lining was to be used.

It went on to say that separate items were set out in the bills for extra costs should the engineer change his mind after the tunnel length had been excavated.

The contractor carried out much more work than was estimated in the bills, but the changes in quantity did not result from a variation order (VO). The engineer granted an extension of time 'to compensate for the extra time required to cope with ground conditions in excavating the [tunnelling] and lining works'.

Mitsui, the contractor, claimed that the increased quantities amounted to variations and it successfully sought extra payment. The employer argued that the engineer had no power to agree or fix adjusted rates, and both parties relied on the contract provisions.

On the agreed facts, the Privy Council ruled in favour of the contractor. The differences between the measured and billed quantities were such as to give jurisdiction to the engineer to either agree or fix a suitable rate, provided he was of the opinion that the nature and amount of the differences was such as to render the billed rate unreasonable or inapplicable.

This depended on the interpretation of clauses 73(2) and 74 of the contract (see the broadly similar provisions in the FIDIC Conditions, 3rd edition, clauses 52(2) and 51(2) of the ICE Conditions, 4th edition).

'The Works' for the purposes of these provisions meant the *original or basic works* and not the works as actually executed, so their Lordships found. The contract contained two definitions of 'the Works', each with different sense.

The Privy Council was able to draw a distinction between the all-embracing definition (including 'additional works') and 'the basic works'.

Clause 73(1) – the variation clause – entitled the engineer to make 'any variation of the form, quality or quantity of the Works', subject to a proviso which appeared to be all-embracing, but which was given the narrowest of interpretations. Lord Bridge of Harwich remarked:

> 'There is much to be said for the view that [the proviso] amounts to a deeming provision whereby any difference between measured quantities for work properly carried out pursuant to the contract and billed quantities is deemed to result from a variation order.'

In fact, although the decision was favourable to the contractor, it seems to fly in the face of earlier rulings and, like the editors of *Building Law Reports*, I think that the case has introduced some uncertainty into what was a well-settled area. 'For judges of first instance and arbitrators', say the editors, '(and indeed some appellate courts) earlier decisions on similar forms of contract are and always have been relevant and in many cases highly persuasive'.

To adopt the approach of the Privy Council can only lead to conflict and uncertainty. The curious feature of this case is that there was a very clear definition of 'additional works' in the contract. That definition was:

> 'All such works which in the opinion of the engineer are of a character similar to those contemplated by the contract and which can be measured and paid for under items in the bills . . .'

The Privy Council appears to have ignored this definition, and introduced a totally unrealistic distinction between 'the Works' (as meaning the basic works) and 'the executed works' – a definition not found in any form of contract.

Civil engineers will study the judgment with interest and, I suspect, with some amazement. It is to be hoped that this will be regarded as a one-off decision. As Lord Bridge said:

> 'Comparison of one contract with another can seldom be a useful aid to construction and may be, as their Lordships think it was in this case, positively misleading'.

I suggest that this is the best way in which to view the *Mitsui* case which should not be held to lay down a general principle. I dislike the Privy Council's cavalier treatment of earlier authorities. What the industry wants is clear and positive guidance on the meaning to be attributed to the terms used in standard-form contracts.

Their Lordships' views in *Mitsui* fly against all the established principles of interpretation and instead use an approach based on 'sensible and businesslike' criteria. One may well ask 'sensible and businesslike to whom'?

Variation problems

The standard form construction contracts empower the architect or engineer to order variations, which cover both extra work and ommissions and sometimes other matters as well. The variations clause is usually drafted in very wide terms (see JCT 80, clause 13) and normally includes the statement that no authorised variation 'shall vitiate this contract'.

There are more misunderstandings and disputes about variations than any other aspect of construction contracts. The starting point is to look at the contract and discover exactly what the contractor agreed to do in the first place, bearing in mind that merely because work is not covered by the bills or specification does not necessarily mean that the contractor is entitled to payment for it as an extra.

In principle, the employer is not bound to pay for 'things that everybody must have understood are to be done but which happen to be omitted from the quantities,' as Mr Justice Channell remarked in the old case of *Patman & Fotheringham Ltd* v. *Pilditch* (1904).

The use of the modern bills of quantities system in fact transfers a substantial element of risk from contractor to employer, and this has led to the system being criticised, notably in Hudson's *Building Contracts*, 10th edition, chapters 5 and 8. But we have to live with the system, and with the problems which it throws up.

The wide definition of the term 'variation' or 'varied work' in the standard forms of contract also gives rise to difficulties. Indeed, it is sometimes suggested that such definition means that the architect can in fact use the variations clause as a means of making fundamental changes in the work.

In fact, this is not so; there is a limit to the 'variations' which can be ordered; the variation must relate to the original work. For example, even where the definition extends to 'omissions', the architect or engineer cannot omit the work so as to give it to another contractor – a point which is illustrated by many cases such as *Commr for Roads* v. *Reed & Stuart Pty Ltd*, (1974) where an attempt to do so was held to amount to a breach of contract by the employer.

Similarly, the contractor cannot be instructed to carry out work which has no connection with the contract by invoking the variations clause. In JCT terms, the extent to which work may be varied must be read in relation to the recitals in the Articles of Agreement stating what 'the Works' are. The contractor cannot be required to do work completely different from the work which he agreed to do. Variation clauses must be read in the light of the circumstances when the parties made the contract.

The point was put shortly by Chief Justice Erle in the shipbuilding case of *Russell* v. *Sa da Bandeira* (1862). The architect or engineer's power to order extras applies only to 'things not specified in, nor fairly comprised within the contract, but cognate to the subject matter and applied to the carrying out of the design'. The contractual definition of 'variation' may also itself limit the architect's powers.

Clause 13.1.1 of JCT 80 refers to 'the addition, omission or substitution of *any* work' – which could scarcely be wider – but clearly this broad phrase must be read by reference to the earlier phrase which mentions 'the alteration or modification of the design, quality or quantity of the works as shown upon the contract drawings and described by or referred to in the contract bills'.

The definition in clause 8.1(d) and (e) of the new ACA Agreement is to the same effect and must be similarly read, despite the earlier reference (clause 1.1) to the contractor's duty to 'comply with and adhere strictly to the architect's instructions issued under this agreement'.

The position seems to be no different under ICE Conditions, or under GC/Works/1, Edition 2, or Edition 3, despite some judicial observations to the contrary in *Sir Lindsay Parkinson & Co. Ltd* v. *Commrs for Works* (1950). However, the ICE reference (clause 51(1)) includes power 'to order any variation that for any other reason shall in his opinion be desirable for the satisfactory completion and functioning of the works'.

The definition of 'variation instruction' in GC/Works/1, Edition 3, is one which 'makes any alteration or addition to or omission from the works or any change in the design, quality or quantity of the works', and in issuing a variation instruction the project manager must decide whether the variation is suitable for valuation on a lump sum basis when he can require the contractor to submit a lump sum quotation.

Under ICE Conditions, it seems that the engineer has no power to order variations in the maintenance period: see clause 49. The ACA Agreement is very clear as to when the architect's power to order variations comes to an end: 'The architect shall have the authority to issue instructions at any time up to the taking-over of the Works' – clause 8.1 – and these include variation instructions. Under JCT terms, the draftsman's original reticence has been clarified by amendment and there is no doubt that the architect can issue variation instructions after the stated completion date is past.

Another problem which arose only under JCT 80 prior to the issue of Amendment No. 4 was in connection with the architect's power (clause 13.1.2) to vary specified obligations and restrictions in the contract bills, e.g. obligations relating to working space, working hours, and so on. The wording of the original clause suggested that the architect could only vary such restrictions as are already placed upon the contractor's methods of working in the contract bills.

The intention clearly was to give the architect power to add *further* restrictions provided they fall within the kinds of restriction listed. For once the Joint Contracts Tribunal heeded the criticism and the clause as amended makes it clear that the architect's power is not confined to varying obligations or restrictions already imposed through the bills but extends to imposing fresh ones of the kinds listed in clauses 13.1.2.1 to 13.1.2.4 and to the contractor's right of reasonable objection.

The ICE variations provisions are much more carefully drafted, but those in the civil engineering industry need not be complacent. Abrahamson's *Engineering Law*, 4th edition pp. 168–196, comments helpfully on some of the potential traps for both employer and contractor. A reading of that textbook dispels some of the common misconceptions.

Variations in construction work are inevitable, but in many cases arise through too much haste to get the project under way, which is a point made by the British Property Federation in its valuable *Manual*.

Value and revalue

The valuation of variations and the amount payable for varied work is one of the commonest sources of dispute in construction and civil engineering projects. The standard form contracts all make similar provisions for valuation, setting out the methods to be adopted in various circumstances, concluding with a 'catch-all' method.

Clause 13.5.1.3, of JCT 80 says:

> 'Where the work is not of a similar character to work included in the contract documents the work shall be valued at fair rates and prices.'

The parallel clause in the ICE Conditions (5th edition) is clause 52(1). This provides:

> 'Where work is not of a similar character or is not executed under similar conditions the rates and prices in the Bill of Quantities shall be used as the basis for valuation so far as may be reasonable, failing which a fair valuation shall be made.'

Varied work is normally paid for promptly. Can the quantity surveyor or engineer later reassess the rate he has fixed in light of subsequent events? And can the arbitrator review such a rate? These are important questions and were answered affirmatively by the High Court in *Mears Construction Ltd* v. *Samuel Williams (Dagenham Docks) Ltd* (1981), where the contract was in the ICE Form (4th edition).

Mears was carrying out piling works, which were to include the use of steel shuttering. The engineer fixed the rates under clause 52, but shortly thereafter it was decided to substitute polypropylene bags for the steel shuttering, which saved Mears a considerable amount of time. Mears claimed the unpaid balance of its final account calculated in accordance with the rates fixed by the engineer. The dock owner disputed the claim on the grounds that it had been fixed on the assumption that steel shuttering was to be used, and that it was not a reasonable rate as polypropylene had been substituted.

The dispute came before His Honour Judge William Stabb QC, the then senior Official Referee, and his ruling is applicable to the ICE Conditions (5th edition) and to the JCT Standard Form of Contract. The ICE Conditions contained an express provision (clause 60(4)) under which the engineer might correct or modify a previous certificate. (Clause 60(7) in the 5th edition.) In fact, he did not appear to have used that power.

Judge Stabb held that this did not matter. He referred to the provision for arbitration (clause 66) and said:

> 'Under clause 66 I consider that the arbitrator has power to open up, review and revise the engineer's certificate which reflects the fixed price when, as here, a dispute has arisen between the employer and the contractor as to the reasonableness of the price . . . I regard the arbitration clause as giving the arbitrator full power to consider and adjudicate upon the reasonableness of the engineer's decision, certificate or valuation, of the variation upon the merits in the light of all the circumstances. I do not believe that he, or the engineer, are to regard themselves as precluded from taking account of subsequent events.'

This statement also applies to valuations of variations under JCT 80 and IFC 84 and to the JCT family of sub-contracts. The wording of the JCT arbitration clause (e.g., JCT 80, clause 41) is equally wide, and Judge Stabb's interpretation is consistent with the nature of interim certificates. The reasonableness of rates is not necessarily to be judged at the time they are fixed and it is permissible to review the matter with hindsight in light of changed circumstances.

Both ICE Conditions and JCT 80 are concerned with 'fair valuation' and 'fair rates and prices'. Said Judge Stabb:

> 'The arbitration clause contains no restrictive provision and therefore . . . gives the arbitrator wide powers which include the power to review and revise, in light of all the circumstances.'

Clause 52 of the ICE Conditions is an extremely complex one and although the current version is not the same as that considered by Judge Stabb, his reasoning clearly applies to it and to analogous JCT clauses. If the original valuation is based on assumptions which do not materialise, then the architect/quantity surveyor or engineer can adjust the valuation later.

Those working under ICE Conditions, 5th edition, are familiar with the complexity of the provisions for the valuation of ordered variations. Clause 52(4) is especially important from the contractor's point of view. It is headed 'Notice of claims', and it lays down a procedure for claims by the contractor for extra payment under *any* clause of the contract and not merely for varied work. The sub-clause provisions as to the giving of notices and provision of information should be noted carefully. Failure by the contractor to comply with these provisions will not necessarily be fatal to his claim. The failure will, however, bar the claim to the extent that the engineer is prevented or substantially prejudiced in investigating it, and such is the importance of clause 52(4) that seven pages are devoted to it in *Civil Engineering Claims* (BSP Professional Books, 1989) which I wrote with Douglas Stephenson.

A case of the unforeseeable

Until recently, the lack of case law on the ICE Conditions of Contract for Civil Engineering Contracts said something for the good sense of the civil engineering industry. Time has changed that, since in the immediate past, there have been many civil engineering disputes before the Courts. One such case is *Hollands Dredging (UK) Ltd* v. *The Dredging and Construction Co. Ltd* (1988), a decision of the Court of Appeal.

The dispute arose out of the construction of a sea outfall pipe and ancillary works at Ardeer, Scotland. The employer was ICI, whose main contract conditions were basically the same as the ICE Conditions. Hollands was the main contractor and sub-contracted with Dredging and Construction Co. Ltd (DCC) in the FCEC blue form of sub-contract. The sub-contracted works included the excavation of a trench landward and seaward of the low water mark, the removal of the spoil, positioning an outfall pipe and then backfilling the trench to seabed level.

The main contract incorporated the minutes of a meeting held in September 1979 between DCC and ICI. The minutes recorded that 'if material had been lost from the dumping ground and there is insufficient to fill in to the original seabed level then ICI will incur no additional cost due to DCC providing additional backfill'. These minutes formed the basis of the dispute.

The loss of excavated material during the backfilling required that extra backfill be obtained by dredging the river estuary and by importing quarried stone. The extra cost was £432,000 above the figure of £90,000 quoted for backfilling in the sub-contract. By agreement, Hollands undertook the further works to finish the construction of the outfall pipe without prejudice to the rights of any party under the main and sub-contracts and without payment from DCC. Both DCC and ICI argued that it was Hollands' duty to obtain all the extra material without further payment.

Certain issues were tried against an assumed factual background. The characteristics of the excavated material made it impossible to backfill the trench to the seabed level using the spoil. This could not have been reasonably foreseen by an experienced contractor.

The trial judge held that neither Hollands nor DCC were entitled to raise claims against ICI in respect of 'adverse conditions' under clause 10 of the blue form and clause 12 of the main contract because the conditions were pre-existing ones and not a post-tendering supervening event. Likewise, Hollands could not raise claims against DCC. The judge also held that no claim could be made under clause 13 of the main contract on the ground of 'impossibility' because the alleged impossibility was not dependent on a supervening event.

Both Hollands and DCC appealed. The Court of Appeal ruled in favour of Hollands and allowed DCC's appeal in part. The main rulings were:

- In interpreting the contract terms, the contracts must be looked at as a whole.
- A method statement, which was an agreed main contract document, was directly relevant to the liabilities arising under the sub-contract: *Yorkshire Water Authority* v. *Sir Alfred McAlpine* (1985).
- Looking at the documents as a whole, including the relevant bill items and the SMM, there was an omission in the specification and bills in respect of the backfill beyond that obtainable from the spoil and additional dredging. Hollands was entitled to have the omission corrected and valued as a variation under clauses 55(2), 51 and 52 of the main contract conditions and clause 10 of the blue form.
- The main contract conditions did not stop DCC or Hollands from applying for a variation under clause 52 in respect of adverse conditions under clause 12. Adverse conditions are not limited to supervening events. However, the sub-contract works were not legally or physically impossible to complete.
- Because the recorded minutes were set down as an agreement, DCC's liability to backfill was greater than Hollands. DCC was liable to dredge other areas of the estuary to make good the material lost

in the general process of excavating. But it was not liable for the imported quarried stone. It was entitled to remedies under clauses 12 and 52 of the main contract similar to those available to Hollands.

This is only a brief summary of a lengthy and important judgment. Its main importance is in relation to clause 11 and 12 claims under ICE Conditions. The only difference between clause 11(1) of the ICE Conditions and the amended version used by the employers is that it qualifies the deemed inspection *'so far as practicable'*.

Had the trial judge's ruling been upheld this would have put many clause 12 claims completely out of court. The trial judge had accepted that DCC could only invoke clauses 12 and 13 successfully if it was able to prove that a sufficient *supervening* event had occurred. The Court of Appeal has said that this is not so.

The Court's attitude towards the SMM is also interesting. The bills and specification were stated to have been prepared in accordance with the SMM which was effectively treated as a code of practice. Yet it was not possible to interpret Hollands' liability as extending to obtaining backfill from any source other than that prescribed in the method statement.

However, the case leaves an important stone unturned. To quote the *Building Law Reports* editors' commentary on the report:

> 'Were a sub-contractor . . . unable to establish the relevant unforeseeability, a clause 12 claim could not be pursued. What is less clear is whether Holland's entitlement to treat the unforeseen shortfall as an *ommission* from the bills . . . remediable under clause 55(2) of the Conditions and to be valued as a variation, would have existed if the shortfall was indeed foreseen.'

For what it is worth, in my view the answer is 'no'.

Variations and their effects: GC/Works/1

Contractors working for the PSA under GC/Works/1, Edition 2, know that varied work is valued in accordance with internal guidance laid down in Technical Instruction QS 5/85, which was drafted following Counsel's opinion that 'expense' in Conditions 9(2) and 53(1) includes finance charges which the contractor proves that he has incurred or is incurring as a consequence of the variation or disruptive event. This opinion was inevitable following the case of *F.G. Minter Ltd* v. *WHTSO* (1980) as amplified and clarified in *Rees & Kirby Ltd* v. *Swansea City Council* (1983).

The PSA's view is that since there is necessarily a delay between the execution of any work and payment for it, the contractor's original bill rates must be taken to include an element for the finance charges to cover that delay. This is undoubtedly the correct principle. In valuing variations under clause 9(1), therefore, there is an appropriate adjustment of any relevant time-related preliminaries which have been separately priced and so far as the variation itself extends those preliminaries.

If the valuation is made under Condition 9(1)(a) it will include the element for finance charges in the original rates, and no further finance charges element is included in the valuation. The same principles are applied under Condition 9(1)(b).

If the variation significantly disrupts the particular varied work then the bill rates and prices no longer apply. The valuation will be done on fair rates and prices (Condition 9(1)(c)) or daywork (Condition 9(1)(d)), whichever is appropriate. Daywork rates are taken to include a finance charges element, and valuation on the fair rates and prices basis also includes an element for them. This deals with any prolongation or disruption directly caused by the variation, but it does not take care of any disruption, etc., caused by the variation to other and unvaried work. It is here that contractors and the PSA disagree.

On a reading of Edition 2, it is plain that the contractor should be able to recover that expense - which will include proven finance charges - under Condition 53, and that is what appears to happen in practice.

There are in fact two types of delay which must be taken into account when valuing the effect of an instruction for modified or additional work. First, there is delay to a critical part of the works which affects overall completion and which results in the contractor retaining his general site overheads for the longer period. Second, there is delay to *part* of the work which does not affect overall completion but causes the contractor to retain some particular item of plant, etc., on site longer than planned.

The bill rates and prices normally form the basis of this valuation and it may be necessary for the contractor to provide a breakdown of the relevant preliminary items in the bill. Using those rates and prices, the quantity surveyor must then value the delay in accordance with the relevant sub-clause of Condition 9(1); normally paragraph (a) will be applicable.

Contractors sometimes allege that the PSA is parsimonious in its approach, but this has not been my experience. It must be remembered that a variation is not a breach of contract and any claim in respect of a variation and its effects must be made under and in accordance with the contract.

Of course, the position may well be different if the quantity surveyor fails to carry out valuations promptly and in accordance with the contract or if the superintending officer fails to certify sums due at the proper time. In that case, I have no doubt that any action might lie for breach of contract and damages would be recoverable, on the basis that a term must be implied into the contract binding the Authority to ensure that the superintending officer and the quantity surveyor properly and timeously perform their duties: *Croudace Ltd* v. *London Borough of Lambeth* (1986) and *Perini Corporation* v. *Commonwealth of Australia* (1969).

In the latter case, a construction contract provided that the Director of Works, a government employee, could grant an extension of time for completion 'if he thinks the cause sufficient [and] for such period as he shall think adequate'. Many of the contractor's applications for extension of time were refused because of departmental policy. The Australian court held, amongst other things, that a term was to be implied into the contract that the government, as employer, would ensure that the Director did his duty as certifier; and damages would be awarded for breach of such an implied term.

Similar argument will, of course, arise under Edition 3 of GC/Works/1, because PSA policy has remained unchanged and prolongation caused by additional or modified work will be valued using the rates and prices in the bills as before. Condition 42 of Edition 3 deals in detail with the valuation of variations, while other instructions are valued under Condition 43. There are some not unimportant differences between the old Condition 9 and the new Condition 43, but the philosophy remains the same, and Condition 42(2) and (5) both contemplate that the cost of disruption and prolongation on both varied and unvaried work must be taken into account.

In truth it has not been my experience that PSA employees act unreasonably. Most of the problems seem to arise because of the contractor's unwillingness to provide a breakdown of his bill rates when requested. In that case, I think, the contractor has only himself to blame for any alleged shortfall, since valuations cannot be done in a vacuum.

Recovering buried problem costs

Although the majority of civil engineering claims flow from changes in the works, other claims under the ICE Conditions, 5th edition, will be associated with ground conditions. These arise under clause 12 of the ICE Conditions, but clauses with similar effect are found in other sets of conditions adopted for civil engineering works.

Under ICE clause 11, the contractor has a duty to examine and inspect the site, and to carry out any other relevant investigation so that he is satisfied with site conditions before submitting his tender. Clause 12 covers *adverse physical conditions* (other than weather) or *artificial obstructions* encountered during the progress of the works. It covers only conditions and obstructions that could not have been foreseen by an experienced contractor at the time of tender.

It is for the contractor to establish that the conditions were not reasonably foreseeable. It is generally accepted that 'a claim is barred only if an experienced contractor could have foreseen a *substantial risk*': Abrahamson, *Engineering Law*, 4th edition, p. 66. The potential scope of the clause is very wide and many civil engineers believe that it is abused. But its purpose and intent is quite clear and it is for the benefit of both employer and contractor.

The employer gains because the contractor does not load his price to take account of remote possibilities. The contractor is not asked to price a risk that he cannot foresee.

The operation of the clause is illustrated by the case of *C.J. Pearce & Co. Ltd* v. *Hereford Corporation* (1968). The contract was for laying new sewers and all incidental works. In the course of the project an old sewer, at least 100 years old, had to be crossed. The contractor knew about it before tender and could expect to meet it anywhere under the embankment where it lay.

In fact the heading for the new sewer met the filled-in original heading for the old sewer under a pathway. Water and silt came into the heading from the old sewer which fractured when the earth surrounding it was disturbed, and as a result the road was blocked. The contractor made a clause 12 claim.

The High Court dismissed the claim. 'An experienced contractor could certainly have foreseen that there might be dangers arising from the driving of the heading immediately under the old sewer,' said Mr Justice Paull. The contractor took the risk upon itself by the terms of the contract.

The contractor must give notice of his claim 'as soon as reasonably possible after the happening of the events giving rise to' it: clause 52(4). The notice must be specific and with it, if practicable, or as soon as possible he must give details of the anticipated effect *and* the measures he proposes to take *and* the extent of the anticipated delay.

The contractor must also be able to show that the condition or obstruction costs more to overcome than could properly have been foreseen.

The physical conditions and obstructions referred to do not

necessarily have to be at the site of the works – the wording is sufficiently broad to cover other physical conditions and thus warrants a claim for payment; for example, a geological fault on adjacent land. Such claims are not uncommon on tunnel projects. The obstructions must be *artificial;* man-made structures and the like. In the context of the clause, all the commentators agree that a physical occurrence is required and not, for example, statutory controls or an obstructive resident engineer!

Despite some statements to the contrary in various procedural handbooks, it is not the *size* of the obstruction which decides whether a claim is to be made under clause 12. No doubt, in practice, small obstructions are sometimes paid under other items in the Bills but, as John Uff has pointed out in his commentary on the ICE Conditions 'the obstructing event may . . . be transient'.

ACA 2 provides optionally for claims for 'adverse ground conditions or artificial obstructions' (clause 2.6) but this is limited specifically to those *at the site.* It is clearly more limited than its ICE counterpart, as to both location and extent. GC/Works 1, Editions 2 and 3, provide a not dissimilar claim, again with strict notice requirements. The intention of all these provisions is the same. They provide an exception to the general rule of law which would otherwise place this risk on the contractor.

Under clause 12, immediately the contractor encounters conditions or obstructions thought to justify a claim, he should do what the clause requires. In practice engineering arbitrators tend to read the clause quite generously, and surprisingly few disputes under clause 12 actually reach the courts. The *Pearce* case is still the main reported English example, and in fact the clause 12 point was only subsidiary to the other points at issue. The contractor had not given notice as required by the clause, and the rectification works had been carried out under an agreement between the contractor's site agent and the city engineer.

The reason for the contractor's failure to give written notice was that, although he had asked who was to bear the cost of the extra work involved, the city engineer replied: 'I'm not talking to you about money. The inquest will come later'. The contractor took this remark as absolving him from giving notice in writing. The judge found that this was not so, and that in any event the problem could have been foreseen by a reasonably experienced contractor.

This spells out the need for contractors to comply exactly with the provisions of the clause, but it should not be used unscrupulously. There is no reverse clause 12, and so if ground conditions turn out to be more favourable than anticipated, the employer has no claw-back.

But most commentators agree that the matter must be looked at in the round and, to quote Abrahamson again, 'a variation of conditions above and below the average could have been foreseen so that the contractor has, overall, no claim'.

Dealing with the unforeseeable

Clause 12 of the ICE Conditions, 5th edition, deals with unexpected adverse physical conditions and artificial obstructions which could not have been reasonably foreseen by an experienced contractor. Under it, the contractor may be entitled both to extra time and to extra payment, although as Abrahamson puts it (*Engineering Law and the ICE Contracts*, 4th edition, p. 65) the clause is 'carefully limited, and it should not be read . . . as giving *carte blanche* to claim or award extra payment whenever the contractor loses out on a calculated risk'.

The High Court decision in *Humber Oils Trustees* v. *Hersent Offshore Ltd* (1981) will be of some help in interpreting this provision, although the case itself turned on the meaning of clause 12 in the 4th edition.

One of the planks in the contractor's argument was that the contract envisaged that the contractor would not be obliged to do that which was physically impossible, and they submitted that:

> 'the philosophy of clause 12 was that the contractor should not bear the risk of unforeseeable physical conditions or artificial obstructions (subject of course to complying with the clause).'

Mr Justice Goff effectively rejected this argument. He did not agree that the terms of the contract should be interpreted or qualified by some nebulous 'philosophy'. The meaning of the contract is to be discerned from the words of the contract itself. He said:

> 'the clause contemplates that the contractor shall not bear the risk of unforeseeable physical conditions or artificial obstructions, but only if the requirement of notice is fulfilled – and so the question remains, what notice is required by the clause? The answer to that question must depend upon the construction of the clause.'

This part of the decision is particularly relevant to clause 12(1) of ICE Conditions, 5th edition, which requires the contractor to give notice to the engineer of the unforeseen conditions, though in rather less stringent terms than its predecessor. This requirement is pursuant to clause 52(4), under which late notice will only prejudice a claim to the extent that it has prevented or substantially prejudiced the engineer from investigating the claim.

In the *Humber Oils Trustees* case the two letters on which the contractors relied as being good notice were in very general terms and, having considered them and the wording of clause 12, the learned judge held that they did not constitute a valid notice.

The first letter referred to 'the difficulty encountered in welding together the free issue pile sections'; the second – sent some seven weeks later – was much more precise, but did not specify the extent of the anticipated delay arising from the physical conditions encountered.

Mr Justice Goff emphasised the importance to the engineer of having the information which is specified in the clause as required to be given in the notice. He said:

> 'The engineer has to assess the strength of the contractor's contention that the conditions encountered are of the relevant type; so he has to have a precise and reliable identification of those conditions. He has to know what is proposed to be done, including what it is likely to cost and how long it is likely to take.'

This is relevant to the 5th edition clause 12, although now under clause 12(2)

> 'following receipt of a notice under sub-clause (1) . . . the engineer may if he thinks fit *inter alia* . . . require the contractor to provide an estimate of the cost of the measures he is taking or is proposing to take . . .'

Only with this information is the engineer really in a position to know what action to take, e.g., to give a suspension or a variation order, and, as the judge observed, these are 'decisions which may be of crucial importance for the future implementation of the contract'.

The purpose of these, or any other notice provisions, is not merely to initiate the claims procedure; it is also to put the engineer on guard so that he may act in the best interests of the employer as well, and take remedial action if necessary.

The learned judge continued:

> 'These considerations point strongly towards compliance by the contractor with the provisions relating to the notice specified . . . and furthermore, since it is common ground that the risk of the effect of the encountered conditions only passes to the employer after the notice is given, these matters also point strongly to the risk

remaining with the contractor until he is able to comply with the requirements relating to notice.

'This will mean that, during any period of assessment of the encountered conditions, and of consideration by the contractor of how to deal with the problem, the risk will remain with the contractor. Of course the clause itself recognises that, in some respects, the information to be given in the notice may not be precise.

'So, even when proposed additional work can be specified, it may not be possible to specify the period of delay; even so, the contractor must give the best estimate he can - that is to say, he can as the clause requires, specify the extent of the anticipated delay, even though this may be, for example, no more than (an estimate), and may have to be modified in the event . . .'

This part of the judgment is of help in interpreting clause 12(1) of ICE Conditions, 5th edition. The procedure specified by the clause must be followed in detail. If he wants to recover money, the contractor must give notice under clause 52(4) 'as soon as reasonably possible after the events giving rise to the claim' and he must specify the physical conditions or artificial obstructions encountered. Either with the notice 'if practicable, or as soon as possible thereafter', the contractor must 'give details of the anticipated effects thereof, the measures he is taking or is proposing to take and the extent of the anticipated delay in or interference with the execution of the Works'.

The decision in *Humber Oils Trustees* v. *Hersent Offshore Ltd* was a victory for commonsense. Contract clauses mean precisely what they say and must be given effect in accordance with their terms.

Site condition problems

In the absence of any specific guarantee or definite representations by the employer about site conditions, the nature of the ground, and so on, the contractor is not entitled to abandon the contract on discovering the nature of the soil. Equally, under the general law, he has no claim for damages against the employer.

The position may, of course, be affected by the express terms of the contract. Clause 2 of GC/Works/1, Edition 2, reiterates the common law rule and places on the contractor the risk that site and allied conditions may turn out more onerous than he expected, though both it and Edition 3 allow claims for certain 'unforeseeable ground conditions'.

The ICE Conditions, 5th edition, clause 11(1), impose on the contractor a duty to satisfy himself 'as to the nature of the ground and subsoil'. In some cases the contractor may have a claim under clause 12 if the actual conditions differ from those foreseen, provided the difficulties specified there 'could not reasonably have been foreseen by an experienced contractor'.

Under JCT terms, the position is more complex, but I adopt the view cogently expressed by John Parris in *The Standard Form of Building Contract JCT 80*, 2nd edition, p 269, where he suggests that the general law is altered 'so far as the JCT Contracts are concerned by what is contained in the SMM' and there is, indeed, case law to support this view, although he does not specifically refer to it.

In *C. Bryant & Son Ltd* v. *Birmingham Hospital Saturday Fund* (1938), Bryant contracted to erect a convalescent home and the contract was in RIBA form with clauses equivalent to JCT 1963 and 1980. The bills of quantities formed part of the contract, clause 11 of which said:

> 'The quality and quantity of the work . . . shall be deemed to be that which is set out in the bills . . . which bills . . . shall be deemed to have been prepared in accordance with the standard method of measurement last before issued by the Chartered Surveyors' Institution.'

The then current SMM required that, where practicable, the nature of the soil should be described, that attention be drawn to any trial holes, and that excavation in rock be given separately.

The bills referred the contractor to the drawings, a block plan and the site to satisfy himself of the local conditions and the full extent and nature of the operations. The architect knew that rock existed on the site, but it was not shown on any of the plans nor referred to in the bills, which contained no separate item for the excavation of rock.

The matter came before Mr Justice Lewis in the form of a special case stated by the arbitrator. He held that Bryant was entitled to treat the excavation in rock as extra, and to be paid the extra cost of the excavation plus a fair profit.

It appears from the arbitrator's findings that he held that it was not reasonable to expect Bryant (or other tenderers) who had seen two trial holes on site (neither of which disclosed the existence of rock) to search the site (which was overgrown with long grass) and find three more trial holes of which they were not aware, but which did show evidence of rock.

There is no material difference between the provisions at issue in the *Bryant* case (where the contract was in RIBA 1931 form) and those of JCT 80. Clause 2.2.2.1 says that the bills are to be prepared in accordance with SMM7, unless otherwise stated, and various provisions in SMM7 require the giving of certain information about ground conditions.

What Dr Parris calls 'the most demanding provision' is D3.2 of the former SMM6 which says that 'if the above information is not available a description of the ground and strata which is to be assumed shall be stated'. In many cases, therefore, the contractor will have a claim and may, under JCT 80, clause 2.2.2.2, treat corrections as a variation for 'any error in description or quantity'.

The position is rather more complex under ICE 5th edition conditions, but the contractor's duty under clause 11 is a qualified one. He is

> 'deemed to have inspected and examined the site ... and to have satisfied himself as to the nature of the ground and subsoil (*so far as is practicable and having taken into account any information ... provided by or on behalf of the employer*) ...' (my emphasis).

All the commentators agree that this has a limiting effect: see, for example, the discussion in Abrahamson, *Engineering Law*, 4th edition, pp. 57–63, where, among other things, it is pointed out that:

(a) Pre-tender investigations may be limited by the time available, and
(b) 'The more extensive be information given by the employer about the site ... the more significance an experienced contractor will reasonably attach to it, and the less it will be practicable for a tenderer to test the information by his own investigations.'

This is a very difficult area and one where it is hard to draw the dividing line. In practice many claims are made in respect of 'adverse physical conditions' under ICE terms.

Quite apart from the contractual situations discussed, there are other remedies which may be available to the contractor in an appropriate situation, notably for misrepresentation. Most of the reported cases come from the Commonwealth, e.g. *Morrison-Knudsen International* v. *Commonwealth of Australia* (1972), where the contractors were misled by site information provided by the employer, which failed to mention that there were cobbles in the subsoil.

Chief Justice Barwick said that the employer's information 'appears to have been the result of much highly technical effort on' its part. It was information which the contractors 'had neither the time nor the opportunity to obtain for themselves'.

The employer may, of course, seek to protect himself by a disclaimer of responsibility – as statutory authorities invariably do when giving information about services – but even if there is a disclaimer, it will not exclude liability for recklessness: see *Pearson & Son Ltd* v. *Dublin Corporation* (1907).

And the courts seem prone to impose liability if it is possible to do so. This is what occurred in *Howard Marine & Dredging Co. Ltd* v. *A. Ogden & Sons (Excavations) Ltd* (1978), where there was an innocent misrepresentation about a barge's deadweight and the makers were held liable.

Chapter 4

Extensions of Time

Power to extend

All the standard forms of building and civil engineering contracts contain clauses providing for the granting of extensions of time linked with a clause for payment of liquidated damages by the contractor in the event of late completion. In the absence of express contractual power to extend the time for completion, the architect or engineer could not do so.

Extension of time provisions and liquidated damages clauses are closely interrelated. If there is power to extend time, and the architect or engineer does not exercise the power on account of delay caused by the employer (or for which the employer is responsible in law), the time for completion becomes 'at large' and the contractor is exonerated from liquidated damages for delay. If time becomes 'at large' then the contractor's obligation is to complete 'within a reasonable time'.

The authority usually quoted for this proposition is the pre-war case of *Miller* v. *London County Council* (1934), a decision of Mr Justice du Parcq. The contract date for completion was 15 November 1931. One clause of the contract provided that:

> 'it shall be lawful for the engineer, if he shall think fit, to grant from time to time and at any time or times by writing under his hand, such extension of time for completion of the work and that either prospectively or retrospectively, and to assign such other time or times for completion as to him may seem reasonable.'

Another clause provided for the payment of liquidated damages for delay at a specified rate. The contractor did not complete the works till 25 July 1932 (some eight months late).

In November 1932 the engineer issued a certificate granting an extension of time to 7 February 1932 and certifying the amount due to the building owner as liquidated damages in respect of the overrun period.

Mr Justice du Parcq ruled that the words 'either prospectively or retrospectively' did not confer on the engineer a right to fix the extension of time *after* the works were completed. They empowered him to wait until the cause of delay had ceased to operate, and then 'retrospectively' with regard to the cause of delay assign to the contractor a new date to work to. As the extension had not been granted in time, no liquidated damages were payable, time for completion having become 'at large'. This case can be criticised on various grounds, but so far as the general rule is concerned it has received the blessing of the Court of Appeal.

This was given in the well-known case of *Peak Construction (Liverpool) Ltd* v. *McKinney Foundations Ltd* (1970), which confirms that if any part of the delay is caused by the employer, no matter how slight, then the liquidated damages clause becomes inoperative unless extensions of time are both grantable *and* granted under the contract. Cautious lawyers will say that whether the completion date is set at large by a delay in granting an extension must depend upon the particular facts of the case and that, indeed, on the present state of the case law is probably the best that can be said.

A New Zealand court has summarised the position in this way:

> 'It must be implicit in the normal extension clause that the contractor is to be informed of his new completion date as soon as reasonably practicable. If the sole cause is the ordering of extra work, then in the normal course extensions should be given at the time of ordering, so that the contractor has a target at which to aim. Where the cause of delays lies beyond the employer, and particularly where its duration is uncertain, then the extension order may be delayed, although even then it would be a reasonable inference to draw from the ordinary extension clause that the extension should be given a reasonable time after the factors which will govern the exercise of the engineer's discretion have been established. Where there are multiple causes of delay, there may be no alternative but to leave the final decision until just before the issue of the final certificate.'

The views thus expressed in *Fernbrook Trading Co. Ltd* v. *Taggart* (1979) are sound commonsense and good contract practice, but in every case the architect or engineer must see exactly what the particular extension of time clause requires him to do. The JCT 80 provisions overcome some of the general difficulties, but clause 25 is by no means perfect, because it does not cover all possible acts of hindrance or prevention for which the employer is responsible.

Under JCT 63, the architect was not required actually to state the new date for completion of the works. Under JCT 80 he is required to fix a new completion date and to inform the contractor of it, although the revised completion date is now a moveable feast because, in certain circumstances, the architect can *reduce* an extension granted.

The general principles are the same under clause 44 of the ICE Conditions, 5th edition, though in the writer's view the provisions there are much more satisfactory than their counterparts in the JCT form. Civil engineering readers will know that 'upon issue of the Certificate of Completion' the engineer must review the contractor's right to an extension and make a final assessment. But he cannot decrease any extension which he has already granted. The crucial factor for the engineer is whether or not the contractor has been held up by a 'delay' under the clause and then whether he is 'fairly' entitled to an extension.

Both under JCT and ICE it is the contractor's *actual* progress and not his planned or hoped-for progress which is relevant. The wording of both contract forms makes this point abundantly clear and this is the answer to those contractors who think that delays are heaven sent and give rise to an automatic extension of time. This cuts both ways since the effect of the delaying event is to be assessed at the time when the work is actually carried out and not when it is programmed to be carried out even if, it seems, the contractor is in culpable delay: *Walter Lawrence & Son Ltd* v. *Commercial Union Properties (UK) Ltd* (1984).

JCT 80 also places on the contractor a duty to 'use constantly his best endeavours to prevent delay *howsoever caused*' as well as to 'do all that may reasonably be required to the satisfaction of the architect to proceed with the works'.

This does not, of course, entitle the contractor to extra payment for any acceleration measures he may take, and he does not have to expend substantial additional money to meet this contractual requirement. There is no authority in JCT 80 or IFC 84 for the architect to order acceleration measures - either with or without payment. The contractor's duty is to prevent delay, so far as he can reasonably do so. This said, extensions of time will forever continue to present problems to architects, engineers and contractors alike.

Extensions and acceleration

Extensions of time under the standard form contracts will always give rise to problems, despite the intention of the clear time-tables laid down in JCT 80 and the ICE Conditions of Contract, 5th edition. The revised extension of time provision in Edition 3 of GC/Works/1 is also clear in its intent, and explicitly recognises that extensions of time may be issued without prior notification from the contractor – as is the case with the ICE provision.

Under clause 36 of GC/Works/1 the project manager must say whether his grant or refusal of an extension of time is an interim or final decision. Interim decisions must be kept under review, and the project manager must 'come to a decision on all outstanding and interim extensions of time within 42 days after completion of the works'.

This explicitly recognises that there can be a retrospective extension of time, which in any event is probably the general rule in most cases. But from all points of view it is desirable to have an extension of time clause which expressly provides for the grant of retrospective extensions of time.

In the New Zealand case of *Fernbrook Trading Co. Ltd* v. *Taggart* (1979), Mr Justice Roper took the view that, under the normal extension of time clause, a retrospective extension of time is only valid in two circumstances:

> '(1) Where the cause of delay lies beyond the employer and particularly where its duration is uncertain . . . although even here it would be a reasonable inference to draw from the normal extension clause that the extension should be given a reasonable time after the factors which will govern the engineer's discretion have been established. (2) Where there are multiple causes of delay there may be no alternative but to leave the final decisions until just before the issue of the final certificate'.

In another New Zealand case – *New Zealand Structures & Investments Ltd* v. *McKenzie* – also heard in 1979, a different judge took the view that under the normal extension of time clause the certifier can grant an extension of time right up until the time he becomes *functus officio*, i.e., devoid of powers, which in most cases will be on the issue of the final certificate. The court said:

> 'In a major contract it is virtually impossible to gauge the effect of any one cause of delay while it is still proceeding, let alone assess the consequences of concurrent or overlapping causes. Finally, any need to have a prompt decision loses some force as a factor in interpreting such a clause, when one considers the normal review and arbitration procedures . . .'

This is a realistic approach, and thus most good contracts provide for both interim and final extensions. The better contracts also have a strict time-table, as in GC/Works/1, Edition 3, which must be observed if the employer's right to liquidated damages is to remain alive. I still consider this to be the situation under JCT 80 despite the apparently contrary comments of Lord Justice Croom-Johnson in *Temloc* v. *Errill Properties Ltd* (1987) where he took the view that the word 'shall' in clause 25.3.3 means 'may'! In modern building practice the principal function of an extension of time clause is surely to enable a date to be fixed for the calculation of liquidated damages for delay, since there must be a date from which liquidated damages can run.

In some cases the argument can be put forward that the contractor is entitled to an acceleration claim where an extension of time is unreasonably delayed. This can be the case under ICE Conditions.

> 'If the contractor is driven to expedite in order to avoid possible liability for damages . . . because the engineer has failed to consider the contractor's right to an extension in good faith at the times at which he is directed to do so by clause 44, then it seems that the contractor may have a claim . . . for damages for breach of contract by the employer by way of failure of the engineer as his agent to administer the contract in accordance with its terms': Abrahamson, *Engineering Law & the ICE Contracts*, 4th edition, p. 372, citing *Morrison-Knudsen Co. Inc.* v. *British Columbia Hydro Authority* (1978).

In principle, there is no reason why such a claim should not be made under JCT 80 or GC/Works/1 for that matter. In all cases the architect or other certifier must administer the contract fairly and in accordance with its terms and certify any extension of time promptly if it is possible to do so. If the certifier delays granting extensions and this causes the contractor to incur acceleration costs, these may be properly recoverable by way of an action for damages.

Thus, in the Australian case of *Perini Corporation Inc.* v. *Commonwealth of Australia* (1969), the engineer repeatedly refused to give a decision on a contractor's applications for extensions of time. The contractor accelerated in order to avoid liquidated damages and to complete on time. He was held entitled to his acceleration costs by way of damages.

It is curious that, as yet, there appear to be no reported English cases on this point, but acceleration claims are frequent in other common law countries. The difficulty would be to establish that the contractor's acceleration of the works was the foreseeable result of the certifier's refusal to award an extension of time, especially as most liquidated damages clauses are for small figures.

Omission instruction confusion

The industry is far from unanimous as to the meaning to be attributed to certain important provisions of JCT 80, clause 25, regarding the architect's power to take account of omission instructions when adjudicating on extensions of time.

The *JCT Guide*, issued with the 1980 form, says:

> 'The architect is permitted after his first decision on extending time to take account of any variations requiring an omission which have been issued since the date of his last decision on extending the completion date,'

thereby implying that he cannot take account of omissions when adjudicating upon the contractor's first extension application.

This view is not, in fact, supported by a reading of the words of the contract itself - and the *Guide* is merely the expression of its author's opinion and is not an aid to interpretation. The difference of opinion centres on clause 25.3.1.4 which says, quite plainly, that in fixing a new completion date the architect is to state to the contractor, 'the extent, if any, to which he has had regard to any (omission instruction), issued since the *fixing* of the previous completion date'. The completion date is defined (clause 1.3) as 'the date for completion as fixed and stated in the appendix or any later date fixed' by the architect under the contract.

This has led to the argument that, contrary to the statement in the *JCT Guide*, the architect's power extends to a variation omitting work that is issued *before* the first extension of time is granted. This argument - with which I am in full agreement - is well rehearsed in John Parris's book. *The Standard Form of Building Contract*, 2nd edition, p. 182, which points out that there is 'always a "previous completion date" from the moment the contract is signed'.

The only criterion for the architect in fixing an extension of time under clause 25.3.1 is that it shall be a 'fair and reasonable' extension of time. In doing so he can take into account any factors which in his view are relevant to the length of extension to be granted and which may mitigate the effects of any delays of which he has been notified by the contractor. These must include the fact that he has omitted work - since, invariably, an omission instruction will save time. Clause 25.3.1.4 merely requires that, if the architect does that when granting an extension of time, he is to inform the contractor accordingly.

Clause 25.3.2 gives him express authority, after he has first granted an extension, to *reduce* that extension

> 'if, in his opinion, the fixing of such earlier completion date is fair and reasonable having regard to (any omission instructions issued), after the last occasion on which (he) made an extension of time'.

The architect does not need express contractual authority to take into account omissions of work when granting extensions of time – they are one of the mitigating factors he must take into account. What he does need is express authority to reduce an extension once granted, and that is what clause 25.3.2 gives him.

The counter argument – which presumably swayed the authors of the *JCT Guide* – is that clause 25.3.1.4 merely requires him to state in his grant of extension of time the extent to which he has taken account of omitted work. This argument supposes that the architect does need authority to take omissions (and other mitigating factors) into account, which is contained in clause 25.3.2, with the result that this power could clearly only be exercised after the first extension of time was granted.

This argument ignores the existence of clause 25.3.1 or at least the whole of its wording. The clause reads (so far as material):

> 'If, in the opinion of the architect, *upon receipt of any notice* . . . under clause 25.2.1.1 . . . the architect shall . . . state . . . the extent, if any to which he has had regard to any instruction . . . requiring the omission of any work . . . issued *since the fixing of the previous completion date.'*

It seems to me that it is a totally incorrect and unsupportable view to take that the architect's power to take omissions into account are governed solely by clause 25.3.2. This very legalistic and strict interpretation seems to fly in the face of the case law on the subject of extensions of time.

Keating's *Building Contracts* (4th edition), p. 351, in commenting upon JCT 63, clause 23, and its reference to 'a fair and reasonable' extension of time exphasises that 'these three words acknowledge that the period of extension can rarely be arrived at by simple processes of arithmetic but has to be the result of a consideration of various factors which may include five listed items'. The extension is to be 'fair and reasonable' for both employer and contractor. The extension of time clause is not there merely for the contractor's benefit. On general principle, omissions can and should be taken into account in assessing what extension of time is to be granted.

Indeed, I think that a correct legal analysis of the whole clause supports the interpretation put forward by Dr Parris, though I am conscious of the fact that quantity surveyors are divided on this point.

The only contractual limitation on the architect's power to take omissions into account is to be found in clause 25.3.6 which says that 'no decision of the architect under clause 25.3 shall fix a completion date earlier than the date for completion stated in the appendix'. That is an overriding provision, no matter how much work is omitted.

Clearly, the architect does need authority to *reduce* an extension once granted - and that is given to him by the contract - but he does not need any authority to take omissions into account in assessing the length of extension which is appropriate in a particular case.

As Keating points out, the process of granting an extension is not arithmetical; the architect must do justice between the parties. Obviously, the contractor cannot be deprived of his original completion period (which is a fundamental basis of the contract), and clause 25.3.6 protects the contractor in that regard. But omission instructions are relevant even on the first grant of extension.

The risk of bad weather

'Exceptionally adverse weather conditions' is one of the grounds for extension of time in JCT 80, clause 25.4.2, and the wording quoted makes it clear that the definition covers extremes of heat and dryness as well as the more normal English weather. (JCT 63 referred to 'exceptionally *inclement* weather' which would not, it is thought, cover drought-like conditions, even though these might have a serious effect on progress.)

GC/Works/1, Edition 2, is more precise. It refers (clause 28(2)(b)) to 'weather conditions which make continuance of work impracticable', while the 5th edition of ICE Conditions is in similar terms to JCT 80. It refers to 'exceptional adverse weather conditions'. In all cases, however, quite unusual severity is required because the contractor is expected to programme his work to take account of the normal weather conditions to be expected at that time of year.

Frost is not exceptional in the United Kingdom in winter months, nor is heavy rain in the spring or autumn. The weather may be exceptional in its duration or intensity but, for it to be a ground for extension of time, it must affect progress of the works. If it does not, then no extension of time can be claimed or granted. Even if the weather is exceptionally adverse for the time of year there is no right to an extension of time if it does not interfere with the works at the particular stage they have reached.

Persistent and torrential rain in London is probably exceptional during the summer months, but not every contractor working there during that period would be entitled to an extension of time. If he was working on the foundations then, I think, in principle he would be so entitled, but not if he was plastering internal partitioning.

How does one determine that weather conditions are 'exceptional'? All the commentators agree that this is to be judged from public records or weather charts covering a long period and relevant to the particular area where the work is to be carried out. The key word is *exceptional* and the matter is fairly put in the Aqua Group's *Contract Administration*, 5th edition, 1981, p. 68:

> 'The word "exceptionally' is clearly the important one . . . and it must be considered according to the time of year and the conditions envisaged in the contract documents. Thus, if it were known at the time that a contract was let that the work was to be carried out during the winter months, and if that work is delayed by a fortnight of snow and frost during January, such a delay could not be regarded as due to exceptionally inclement weather. If, however, such work was held up by a continuous period of snow and frost, from January until the end of March, an extension of time would clearly be justified under this clause.'

Location of the work is important, as the authors point out, and they add that 'in certain circumstances such delays may be avoided by providing *in the contract* for additional protection and even temporary heating arrangements'. Under JCT terms, of course, the contractor is under an obligation to take all practicable steps to avoid or reduce delay and also to do all that may reasonably be required to the satisfaction of the architect to proceed with the works, and this proviso qualifies the right to extension of time.

Some architects have argued that, as a result, the contractor can be required to expend money. This is not so and hence, presumably, the Aqua Group's reference to making provision in the contract for temporary works, etc. The contractor need take only ordinary measures to proceed or, as I have put it elsewhere, his obligation is 'to show willing'.

Generally, of course, there are concurrent or parallel causes of delay some of which, like exceptionally adverse weather, merely give rise to an extension of time and others, like late information, which may give rise also to a financial claim. In such circumstances the contractor often argues for an extension of time on the ground which also gives rise – independently – to a loss and/or expense claim and is not at all happy if the architect opts for adverse weather as the ground for extension.

This attitude displays a fundamental misconception about the contractual position. Disruption costs arising from late information are clearly reimbursable either under the express contract terms or, alternatively, at common law. Where there are parallel causes such as those I have mentioned, the question for the architect is simply, 'Is the progress of the works materially affected by the late information'? If the answer to that question is 'no' because no work could have proceeded even had the information been available on time, i.e., because of the inclement weather, there is no claim for loss and/or expense.

This view is not popular with contractors, but it seems to me (and to others) that this is what the contract says. There is no case law on the topic - though it derives some support from observations of Mr Justice Megarry in *Hounslow Borough Council* v. *Twickenham Garden Developments Ltd* (1971), as well as the major textbooks. At the very best, it is a grey area. Under JCT extension of time clauses it is the effect of the relevant event upon the completion date which is critical. In contrast, under JCT 80, clause 26, it is the effect upon *progress* which matters, and if the late information has no effect upon the progress of the works (which does not necessarily delay the completion date) there can be no entitlement to loss and/or expense.

Some writers argue, in fact, that JCT terms do not recognise disruption claims as opposed to prolongation claims, and would thus disallow compensation for loss of productivity: see, e.g., Parris, *The Standard Form of Building Contract*, 2nd edition, p. 167, but most authorities say that a loss of productivity claim may be properly made.

Statutory commotion

Clause 25.4.4 of JCT 80 provides one ground on which the contractor may be entitled to extension of time. It refers to

> 'civil commotion, local combination of workmen, strike or lock-out affecting any of the trades employed upon the works or any of the trades engaged in the preparation, manufacture or transportation of any of the goods or materials required for the works.'

The language used is both archaic and obscure.

An interesting point revolving around this provision, in its JCT 1963 form, came before Mr Justice McNeill in *Boskalis Westminster Construction Ltd* v. *Liverpool City Council* (1983), by way of an application for leave to appeal on a point of law under sl(2) of the Arbitration Act 1979, arising out of an arbitrator's award.

Boskalis contracted in JCT 63 form with the council. There were delays in completion and disputes were referred to arbitration. The parties were agreed that works by statutory undertakers did not form part of the contract, statutory undertakers being directly employed by the council.

The question before the arbitrator was a simple one: is 'delay to the completion of the works by reason of strikes affecting workmen employed by . . . the statutory undertakers' a matter falling within clause 23(d)?

The arbitrator answered this question 'no', and made an award in favour of the contractor in respect of loss and/or expense it had suffered as a result. His award was under what is now clause 26 of JCT 80, and he ruled on the evidence 'that a strike by workers employed by statutory undertakers directly engaged by the [council] to execute work not forming part of the works' was not covered by clause 25.4.4.

The position would have been different had the statutory undertakers' work formed part of the contract, as the same arbitrator found and held in *Henry Boot Construction Ltd* v. *Central Lancashire New Town Development Corporation* (1980) where sums relating to the work of statutory undertakers had been included in the contract sum.

Mr Justice McNeill refused the council leave to appeal, and in so doing made some interesting observations. He pointed out that if the JCT contract had been drafted to provide that any effect upon the works by any strike would disentitle the contractor to reimbursement under clause 24, it would have been simple to say so. He said:

> 'Here, the question which the arbitrator was asked to decide was whether the delay by reason of strikes affecting workmen employed by . . . the statutory undertakers came within clause 23(d).'

He regarded the question as 'unhappily phrased'.

The arbitrator had not been asked to say whether there was delay by reason of strike affecting any of the trades employed upon the works'. In the context of the concession that was made . . . upon the question put before him . . . I do not think that the answer can be faulted.' The judge regarded the question as a 'one off' situation on the facts, and consequently he did not feel that there was any substantial point involved.

In principle this decision is correct on the facts, but in my view *Boskalis Westminster Construction Ltd* v. *Liverpool City Council* is not a definitive interpretation of what is now clause 25.4.4. It is at least arguable that a strike of, for example, the water authority's workmen engaged in laying mains would be a strike 'affecting any of the trades employed upon the works'. Judge Fay's decision in the *Henry Boot* case is still relevant as to the meaning of the words 'not forming part of this contract,' which appear in clauses 25.4.8, 26.2.4 and 29 of JCT 80, and that judgment still repays careful study.

More generally, clause 25.4.4 raises problems of interpretation in today's context. What, for example, is 'civil commotion'? It has been said that it is a term used 'to indicate a stage between a riot and civil war' and there must be an element 'of turbulence or tumult': Lord Justice Luxmoore in *Levy* v. *Assicurazioni Generali* (1940). In a contract carried out in Northern Ireland there must be daily examples, one assumes.

'Local combination of workmen' has a Victorian ring about it; but 'strike or lock-out' is an up-to-date phrase. Both official and unofficial strikes are covered, but the term does not extend to cover 'working to rule' or other obstructive activities falling short of a strike. Perhaps today such localised action *might* conceivably be held to amount to a 'local combination of workmen'.

The real point at issue in all these cases is, of course, whether the contractor is entitled to reimbursement of any loss or expense which he suffers or incurs as a result of the delay. Clause 25.4.4 is not paralleled in clause 26: a strike, etc., may entitle the contractor to an extension of time, but he does not get any money, for clause 25.4.4 is one of those events regarded by the contract as being outside the control of either party.

If, on the other hand, the contractor can bring his claim under clause 26.2.4, for example, as being originated by 'execution of work not forming part of this contract by the employer himself or by persons employed or otherwise engaged by the employer', he has a money claim as well.

In 1980, Judge Fay summarised the relationship between what are now JCT clauses 25 and 26 in this way:

> 'The broad scheme of these provisions is plain. There are cases where the loss should be shared, and there are cases where it should be wholly borne by the employer. There are also cases which do not fall within either of these two conditions and which are the fault of the contractor, where the loss of both parties is wholly borne by the contractor. But in the cases where the fault is not that of the contractor the scheme clearly is that in certain cases the loss is to be shared: the loss lies where it falls. But in other cases the employer has to compensate the contractor in respect of the delay . . . [which]

should . . . clearly be composed of cases where there is fault upon the employer or fault for which the employer can be said to bear some responsibility.'

That was the real point between the parties in *Boskalis*; on the facts the arbitrator gave the right and just answer.

Timely questions under ICE Minor Works Form

In contrast with the position under claue 44(1) of the ICE Conditions of Contract, 5th edition, the test to be applied in granting an extension of time under clause 4.4 of the ICE Minor Works Conditions is whether 'the progress of the works or any part thereof shall be delayed' for any of the specified reasons. Under the ICE Conditions the test is whether the event entitles the contractor to an extension of time.

The authors of the form included as a sweeping-up ground the almost traditional occurence of 'other special circumstances of any kind whatsoever *outside the control of the contractor*'. The italicised words do not appear in the ICE Conditions themselves, but despite their inclusion the position appears to be the same as under clause 44(1).

This is well summarised in John Uff's commentary on those words in Keating's *Building Contracts*, 4th edition, p. 493:

> 'Such general words will not be construed as entitling the engineer to grant an extension of time in respect of delay due to the employer's default. Unless such default falls within an express ground for extension, such delay will invalidate the liquidated damages clause . . .'

It would have certainly been better to have used words expressly covering any act, default, negligence or omission of the employer and those for whom he is responsible in law as a ground for extension of time. This criticism apart, the provision for extension of time is superficially unexceptionable. Obviously the sponsoring bodies think this is so since there is no comment on clause 4.4 in the *Guidance Note*.

Eight grounds are specified as triggering off a claim for an extension of the period for completion. They are:

- Variations, suspension orders and instructions changing the intended sequence of the works;
- Carrying out tests or investigations where the result does not disclose non-compliance with the contract;

- Encountering artificial obstructions or physical conditions;
- Delay in receipt by the contractor of necessary drawings, instructions or other information;
- Failure of the employer to give adequate access or possession;
- Exceptional adverse weather;
- Delayed receipt of materials to be provided to the contractor by the employer;
- Other special circumstances outside the contractor's control.

If progress of the works is delayed because of one or more of these events, the engineer is bound

> 'upon a written request by the contractor *promptly* by notice in writing to grant such extension of the period for completion . . . as may in his opinion be reasonable'.

The contractor's notice is not, it is thought, a condition precedent to the grant of an extension of time. If it were to be so regarded, one of the objectives of the clause would be defeated, namely to preserve the employer's right to liquidated damages for the late completion.

The wording of the clause also seems to prohibit the not uncommon practice of postponing consideration of extensions of time until the end of the contract. The engineer is also under a duty to subject 'the extended period or periods for completion . . . to regular review', with the proviso that no review can decrease any extension of time which he has already granted.

Nothing is said about the circumstances which the engineer must take into account in assessing what extensions to grant. There are no defined criteria except that the progress of the whole or part of the works is delayed. Clearly, however, the engineer must have regard to such things as omissions, and the contractor's clause 4.3 programme will be a useful guideline.

The lawyers will certainly not welcome the lack of precision, but I suspect that practical engineers and contractors will do so.

The engineer's decisions on extensions of time are reviewable by the arbitrator, who has 'full power to open up, review and revise any decision, instruction, direction, certificate or valuation of the engineer': clause 11.7. Civil engineering arbitrators in practice do not seem to take an over-legalistic view of extension of time clauses. Courts, however, do so.

Liquidated damages are linked with extensions of time and are covered by clause 4.6. The grant of an extension of time under clause 4.4 postpones the date from which liquidated damages start to bite.

That is, indeed, the only effect of grant of an extension of time under the Conditions.

Under this form liquidated damages are subject to a limit stated in the Appendix which it is suggested should not exceed 10% of the final contract value as estimated at the tendering stage. This, says the *Guidance Note*, 'should be taken into account when assessing the daily or weekly rate' of liquidated damages.

Although the possibility of a daily rate is suggested, it would be better to stick to a weekly rate in the Appendix because clause 4.6 refers to liquidated damages 'for every week' and provides for their *pro rata* recovery for part of a week.

There is no express provision for the deduction of liquidated damages, but merely a statement that 'the contractor shall be liable to the employer' for liquidated damages. The employer will therefore have to rely on his common law and equitable rights of set-off.

Clauses 4.4 and, to a greater extent, 4.6 are really the only disappointing part of what is otherwise a very good contract. There has been much case law in the area, and it is a common source of dispute.

It is a pity, therefore, that these two important provisions were not drafted in a better form. Much is left unsaid, and the draftsman should have been left a freer hand than he was.

The problem of delay

Contractors often complain that architects and engineers are parsimonious in the granting of extensions of time. That would certainly seem to have been the intention of the draftsman of JCT 80 which introduced the very valuable architect's review of extensions of time in clause 25.3.3. The architect must carry out the review in the light of any causes of delay – called somewhat inelegantly 'relevant events' – whether or not they have been notified to him by the contractor.

The object of this is, of course, to preserve the employer's right to liquidated damages. If the architect fails to carry out his review and grant any appropriate extension of time, then the contract time will become 'at large'. There is then no date from which liquidated damages can run, and hence none will be recoverable. The employer would be left to claim general damages at common law – on the basis of his proven losses. If time is 'at large' the contract completion date ceases to be applicable and the contractor's obligation becomes one to complete 'within a reasonable time'.

What is a reasonable time will depend on all the facts and

circumstances and I discuss the point further on p. 77. As a rough guide, it is probably the original contract period plus a period equivalent to any extension that ought reasonably to have been granted had the architect acted correctly.

The problem was more acute for those working under JCT 63 contracts because of its totally defective extension of time clause. That provision – clause 23 – did not contain a reviewing power, and also lacked a number of grounds now found in JCT 80, clause 25, e.g. 'the supply by the employer of materials and goods which the employer has agreed to provide for the works or the failure' so to supply. Moreover, the architect operating under JCT 63 had no power to grant an extension of time once the due date for completion has passed.

This is of great importance in respect of events giving rise to an extension of time which are employer's fault or responsibility, such as late information and so on.

It is quite clear and settled law – despite assertions to the contrary – that if the employer is wholly or partly responsible for the delay or if an extension is not made at the right time, then the liquidated damages clause becomes inoperative and time becomes at large. The authority for this proposition is the well-known case of *Peak Construction (Liverpool) Ltd* v. *McKinney Foundations Ltd* (1970). Although that case involved an in-house contract form, the principle applies equally to JCT 63 or 80. The headnote is quite clear as is the holding:

> 'If the employer is in any way responsible for the failure to achieve the completion date, he can recover no liquidated damages at all and is left to prove such general damages as he may have suffered.'

More recently, in *Rapid Building Group Ltd* v. *Ealing Family Housing Association Ltd* (1985), the Court of Appeal upheld the *Peak* case in the context of a JCT 63 contract and ruled that its authority was binding. Lord Justice Stephenson said:

> '. . . if the employer is responsible for any delay which does not fall within the *de minimis* rule, it cannot be reasonable for him to have the works completed on the completion date. Whatever the reasoning underlying the decision of this court it binds us . . .'

The *Rapid Building* case is interesting in another respect and a paragraph in the judgment of Lord Justice Lloyd is worth quoting at length:

> 'Like Lord Justice Phillimore in [the *Peak* case] I was somewhat

startled to be told in the course of argument that if any part of the delay was caused by the employer, no matter how slight, then the liquidated damages clause in the contract becomes inoperative. I can well understand how that must necessarily be so in a case where the delay is indivisible and there is dispute as to the extent of the employer's responsibility for that delay. *But where there are, as it were, two separate and distinct periods of delay with two separate causes, and where the dispute relates only to one of those two causes then it would seem to me just and convenient that the employer should be able to claim liquidated damages in relation to the other period.'*

This statement was not part of the holding in *Rapid Building* but, coming as it does from a Lord Justice of Appeal must carry a good deal of weight and sooner or later it will be seized upon and argued out.

In principle, once the employer's right to liquidated damages has gone it cannot be revived. He is left to pursue his other remedy and claim unliquidated damages. Presumably what Lord Justice Lloyd had in mind was the situation where there is a neutral event which is the fault of neither party, such as exceptionally inclement weather, and an event which is employer's fault. Even in that case I cannot, with respect, see that it is 'just and convenient' to allow the employer part liquidated damages. But perhaps his lordship had something else in mind, e.g., where the contractor is himself in default and then there is delay caused by the employer.

However, even then the sensible solution would be for the employer to pursue a damages claim at common law, even if he is then put to proof of loss. For the moment, the position is as stated in *Peak*. If completion by the specified date is prevented, wholly or partly, by the fault of the employer, he can recover no liquidated damages unless there is an extension of time clause providing for an extension of time on that ground, and the architect or engineer grants an extension of time as specified in the contract: *Percy Bilton Ltd* v. *Greater London Council* (1981), a decision of the House of Lords.

Time at large

All the standard form contracts contain clauses providing for the date by which (or period within which) the project must be completed. Thus, JCT 80, clause 23.1.1 provides:

'On the Date for Possession possession of the site shall be given to the contractor who shall thereupon begin the works, regularly and

> diligently proceed with the same and shall complete the same on or before the completion date.'

The completion date is defined in clause 1.3 as 'the Date for Completion as fixed and stated in the Appendix or any date fixed under . . . clause 25 . . .'. The Appendix sets out the date which, subject to any extensions of time, is the contractual 'Date for Completion'.

A risk associated with a fixed completion date is that the employer may by some act or default make the completion date inapplicable, with the result that the contractor is no longer bound to complete by the stated date, but only 'within a reasonable time'. This is the concept of 'time at large', and in practical terms it has the important consequence that, since there is no fixed date for completion, the employer loses his right to liquidated damages.

JCT 80 and other standard contracts address this problem by allowing the grant of an extension of time to cover acts or defaults of the employer which cause delay and which, in the absence of an extension of time, would result in time being at large. The contract mechanisms protect the employer from this risk.

However, the problem is not so easily resolved because it may be that the extension of time clause is not sufficiently widely drawn to cover all acts or defaults for which the employer is responsible. There is also the danger that the architect (or engineer) may fail properly to grant an extension when one is due.

The classic example, of course, would be the employer's failure to give possession of the site as agreed; in the absence of power to extend time for such failure, there is no doubt that time would be at large. This was a problem under JCT 63 and was the case under JCT 80 prior to July 1987 when clause 23.1.2 was inserted to provide:

> 'Where clause 23.1.2 is stated in the Appendix to apply the employer may defer the giving of possession for a period not exceeding six weeks or such lesser period stated in the Appendix calculated from the Date of Possession.'

However, this does not necessarily solve the problem in every case, and Martin King of London solicitors Denton Hall Burgin & Warrens suggests that the possession clause should be amended to provide:

> 'The employer shall on the date for possession give to the contractor possession of so much of the site as may be necessary to enable the contractor to begin the works, and the contractor shall proceed with the same regularly and diligently. The employer will, from time to

time as the works proceed, give to the contractor possession of such further portions of the site as may be necessary to enable the contractor to complete the works on or before the completion date.'

If time does become 'at large', the contractor's obligation is to complete within a reasonable time. What is a reasonable time is a question of fact: *Fisher* v. *Ford* (1840). Calculating a reasonable time is not an easy matter and, as Emden's *Building Contracts*, 8th edition, vol. 1, p. 177, puts it:

> 'Where a reasonable time for completion becomes substituted for a time specified in the contract . . . then in order to ascertain what is a reasonable time, the whole circumstances must be taken into consideration and not merely those existing at the time of the making of the contract.'

The illustration quoted is *Charles Rickards Ltd* v. *Oppenheim* (1950), where Rickards agreed to supply a Rolls Royce motor car chassis and to build a body on it within seven months. They failed to complete the work by the agreed delivery date, but Oppenheim waived the original delivery date and new dates were promised and accepted by him. Eventually, Oppenheim gave written notice to Rickards stating that unless he received the car by a firm date, four weeks away, he would not accept it. The car was not delivered within the time specified and was not completed until some months later when Oppenheim refused to accept it.

The Court of Appeal held that he was justified in doing so. After waiving the initial stipulation as to time, Oppenheim was entitled to give reasonable notice making time of the essence again, and on the facts the notice was reasonable.

In a building context it is clear that the same principles apply. If for some reason time under a building contract becomes 'at large' the employer can give the contractor reasonable notice to complete within a fixed reasonable time, thus making time of the essence again: *Taylor* v. *Brown* (1839). However, if the contractor does not complete by the new date, the employer's right to liquidated damages does not revive, and he would be left to pursue his remedy of general damages at common law.

The problem of time becoming at large does not appear to have been thought out by the Joint Contracts Tribunal. The JCT 80 extension of time provision (clause 25) would be better by far if it followed the policy of ACA2 and had as a sweeping-up ground for extension of time 'any act, instruction, default or omission of the employer, or of the architect on his behalf, whether authorised by or in breach of this Agreement': ACA2, clause 11.5.

Chapter 5

Liquidated Damages

Problems with liquidated damages

All the standard form building contracts contain a provision for the contractor to pay liquidated and ascertained damages should he fail to complete the works on time. The object of such a clause is to fix an amount which is a genuine pre-estimate of the employer's loss in the event of late completion or, in practice, a lesser sum.

For a liquidated damages clause to operate successfully there must be a simple contractual machinery to produce a readily calculable mathematical result, and this is provided by all the JCT standard contracts, although recent case law suggests that the operation of the machinery is often misunderstood by architects and contractors alike. Although the legal principles involved are well settled, there have been recent case law developments which show the confusion surrounding the subject and leave at least one aspect of the law in a state of flux.

The important point about liquidated damages is that they are recoverable without the employer having to prove loss and it is irrelevant if in the event there is no loss. This proposition has long been established law (*Clydebank Engineering Co. Ltd* v. *Castadena* (1905)) but is often challenged by contractors. A recent example is *BFI Group of Companies Ltd* v. *DCB Integration Systems Ltd* (1987) which arose out of an alteration contract of a transport depot carried out under the JCT Agreement for Minor Works 1980. There was a six week delay in completing two of six loading bays, but because the employers had to fit the building out after completion, this did not cause them any loss of revenue. Judge John Davies QC, Official Referee, held that it was quite irrelevant to consider whether there was any loss in fact. He said:

> 'Much as I instinctively dislike provisions for liquidated damages, a provision of this sort was one which automatically came into play once the event happened. There is no question here of it being a penalty, as the arbitrator had himself decided that it was not. It was

an entire contract for six bays, two of which were not ready for six weeks, and the damages were pre-estimated in the contract.'

Liquidated damages are often referred to erroneously as 'a penalty' and some employers and their advisers think of them as a stick with which to beat the contractor. This temptation should be avoided since any tendency to increase the weight of the stick can result in the sum inserted in the contract being classed as an unenforceable penalty.

The distinction between a penalty and liquidated and ascertained damages does not depend on how the parties describe the provision but on its *purpose*. In practice, the distinction may be difficult to draw on the facts, but if the provision is held to be a penalty it will be unenforceable.

To avoid being a penalty the figure stipulated for must be a fixed and agreed sum which is a genuine pre-estimate of the likely loss to the employer or a lesser sum. The genuineness of the estimate is to be judged at the date the contract is entered into and not at the time of the breach. A penalty, in contrast, is an extravagant sum of money inserted to coerce the contractor to performance: see *Dunlop Pneumatic Tyre Co. Ltd* v. *New Garage & Motor Co. Ltd* (1915), for the classic guidelines laid down by Lord Dunedin. If a figure is held to be a penalty, then the employer is entitled to sue for general (unliquidated) damages on the basis of his proven losses, although possibly subject to there being a ceiling on the amount claimed equivalent to the failed clause.

The fact that it is difficult to make an accurate estimate of the financial consequences of the breach does not prevent the figure being a genuine pre-estimate. This is of particular importance in public works contracts where estimating the likely loss is inherently difficult.

In fact, there have been few building industry cases where a provision has been held to be an invalid penalty because of its amount, but in other areas of the law the application of the distinction between liquidated damages and penalties has given rise to considerable difficulties. In practice, in the industry the amounts agreed as liquidated damages are relatively small in relation to the employer's likely loss.

There have been recent construction cases where a figure in a construction contract was held to be a penalty. In the first – *Stanor Electric Ltd* v. *R. Mansell Ltd* (1988) – electrical sub-contractors were required to carry out works at two houses in London, the sub-contract providing for liquidated damages 'and/or damages for non-completion and/or delay' at the rate of £5000 per week. According to the documents before the High Court, the sub-contractor 'failed to complete one of the two houses . . . by the contractual date and delayed by two weeks thereby incurring liquidated and ascertained damages of £5000 (i.e., one half of two weeks at £5000 per week)'. The High Court held that the clause was 'self-evidently a penalty'. What was a reasonable sum for failure to complete the whole works, i.e., the two houses, could not possibly be a genuine pre-estimate in respect of only one house.

The second and better-known case is *Bramhall & Ogden Ltd* v. *Sheffield City Council* (1983), where the contract was in JCT 63 form and was for the erection of 123 dwellings and ancillary works. The date for completion was stated in the appendix as 6 December 1976 and the appendix entry for liquidated damages was filled in: 'at the rate of £20 per week for each uncompleted dwelling'. No sectional completion supplement was entered into.

The architect granted extensions of time up to 4 May 1977. Thereafter, as houses were completed, they were taken over by the employer, the last on 29 November 1977. The employer retained £26 150 (i.e., 123 × £20) as liquidated damages for the houses completed between 4 May and 29 November 1977.

The High Court held that the council were not entitled to liquidated damages. The contract made no provision for sectional completion. Since possession of completed houses was taken by the council from time to time with the contractor's consent, JCT 63, clause 16(e) (now JCT 80, clause 18) applied. The contract provided for liquidated damages on the contractor's failure to complete 'the works', i.e., all the dwellings and not just some of them, and clause 16 could not operate where the rate in the Appendix was expressed as it was. The clause failed because both the damages and the reduction provisions (clause 16(e)) were expressed by reference to the whole works while the liquidated damages were expressed by reference to a single dwelling.

Calculating liquidated damages is never easy. The essential point is that a genuine pre-estimate should be made and this involves an actual calculation using verifiable data. This should be retained so that the calculation can be sustained if the figure is later challenged. It is sufficient to make a genuine and informed estimate from available data and it matters not that the estimate turns out to be a poor one in fact. But figures should not be plucked out of the air.

Where, as in *Bramall & Ogden*, the works consist of a number of individual units, the figure must not be expressed as a global sum for the whole contract as it will then be an unenforceable penalty. If the contract is for 100 dwellings, what is a genuine pre-estimate for all the houses together cannot possibly be a genuine pre-estimate for, say, 10 houses which are late. There must therefore be some contractual mechanism for the proportionate reduction for a lesser number of houses than the total. JCT 80 provides for this but, as we have seen, the mechanism may be difficult to apply in practice.

Other recent legal decisions show the complexity of this area. For example, in *Temloc Ltd* v. *Errill Properties Ltd* (1987), there was a JCT 80 contract for the construction of a large shopping development at Plympton, Devon. The contract sum was £840,000. Clause 24 (the liquidated damages clause) was left in the printed contract as executed by the parties but the relevant appendix entry was filled in as '£NIL'. The contract overran by some five weeks, and as a result prospective tenants of the development brought actions against the developer who in turn sought to claim over against the contractor.

The Court of Appeal held:

- The effect of '£NIL' was not that clause 24 was ineffective or to be disregarded, but was left to apply in the negative way expressed in the appendix. On the proper interpretation of the contract, the parties had agreed that there should be no damages for late completion.
- Liquidated damages were exhaustive of the employer's remedies for the breach of late completion and thus he was not entitled to claim general (unliquidated) damages on proof of loss. Any claim for such damages would have to be based on an implied term - one written into the contract on the basis of the parties' presumed intentions - but because of the clear provisions of clause 24, no such term could be implied.

This decision has attracted some criticism, but is clearly binding authority, and although it does not decide the point it may well be the start of a trend against being able to claim general damages where the liquidated damages clause has failed. In the earlier case of *The Rapid Building Group Ltd* v. *Ealing Family Housing Association Ltd* (1985), the Court of Appeal stated that 'where the claim for liquidated damages has been lost or has gone . . . the defendants are not precluded from pursuing their counterclaim for unliquidated damages'. Their lordships left open the point as to whether the amount of those damages is limited to the amount of the defunct liquidated damages. It is certainly a sustainable view that any claim for unliquidated damages has on it a ceiling equal to the amount of the failed liquidated damages provision. If general damages in excess of the stipulated amount are recoverable, that would be an added bonus for the employer.

Clearly, architects must take care when treading through this minefield, and it must be said that both the law and the practice are in a state of flux. In my experience, liquidated damages are generally recovered successfully only in very straightforward cases and more and more contractors seek to challenge the validity of such provisions. If a dispute arises and is pursued to arbitration or litigation, the employer may be able to justify the deduction of liquidated damages or, if they have failed for some reason or other, establish a right to general damages at common law. But such disputes are costly and time-consuming and if, as is usually the case, the contractor has a claim for direct loss and/or expense, the dispute is often compromised on terms that the employer foregoes his right to liquidated and ascertained damages.

Operating JCT 80, clause 24

Some common misunderstandings about the JCT 80 provisions for liquidated damages and extensions of time were cleared up by the important decision of His Honour Judge John Newey QC in *A. Bell & Son (Paddington) Ltd* v. *CBF Residential Care and Housing Association Ltd* (1989). The amount at issue - £4900 - was trivial, but there was an important practical point involved, namely the correct interpretation of clause 24.1. This provides:

> 'If the contractor fails to complete the works by the completion date then the architect shall issue a certificate to that effect'.

The architect's certificate of non-completion is a condition precedent to the employer's right to liquidated damages, and 'the completion date' is the date stated in the appendix or as fixed under the extension of time clause.

Judge Newey had no doubt about the correct interpretation:

> 'When a new completion date is fixed, if the contractor has not completed by it, a certificate to that effect must be issued and it is irrelevant whether a certificate has been issued in relation to an earlier, now superseded, completion date. [This interpretation] accords with the setting of the contract: contractors and employers using it need above all certainty and the issue of a fresh certificate will provide it.'

The learned judge went on to consider clause 24.2.1 which introduces a further condition precedent to the deduction of liquidated damages, namely that the employer must give written notice to the contractor of his intention to claim or deduct liquidated damages and he must do this before the issue of the final certificate. The contractor is only bound to pay or allow 'to the employer the whole or part of such [liquidated damages] as may be specified in writing by the employer'.

Since the giving of notice is made subject to the issue of a certificate of non-completion, if the certificate is superseded, then logically the notice should fall with it. Judge Newey thought that here

> 'the setting of the contract may point in the opposite direction, for once an employer has informed a contractor of his intention to recover liquidated damages he is unlikely to change his mind.'

But, he ruled, 'once again certainty is the greatest need and . . . if a new completion date is fixed any notice given by the employer before it is at an end'.

This sensible ruling is in sharp contrast to some of the more recent decisions on liquidated damages given in the more rarified atmosphere of the Court of Appeal, which does not include any former official referees whose day-to-day task is grappling with construction industry problems, and construction law involves a unique combination of both law and practice.

The facts of *Bell* v. *CBF Housing Association* are not atypical. The JCT contract was one for the extension of a house in The Bishops Avenue, London N2, and the contract sum was £371,188. The date for possession of the site was 28 May 1985 and the original completion date was 28 February 1986. Liquidated damages were at the rate of £700 a week.

Bell did not complete by the latter date and gave notice of delay under clause 25. The architect awarded an extension of time so that the new completion date was 25 March. The plaintiffs did not complete by the new date and the architect issued a clause 24.1 certificate. CBF wrote to Bell on 3 April 1986 informing him of its intention to deduct liquidated damages.

On 3 June 1986 the architect awarded Bell a further extension of time, fixed 14 April 1986 as the new completion date and granted a further week's extension on 23 June so that the completion date became 21 April. Bell achieved practical completion on 18 July 1986.

More than a year later – on 3 December, 1987 – the architect granted another extension so that the completion date became 20 May 1986. (The architect ought to have given his final decision under clause 25.3.3 within 12 weeks of practical completion and why he failed to do so is not clear from the judgment, though the extraordinary interpretation put on clause 25.3.3 by the Court of Appeal in *Temloc Ltd*

v. *Errill Properties Ltd* (1987) seems to endorse this common delaying tactic of architects.) The final certificate was issued on 25 February 1988.

CBF paid Bell the amount of the agreed final account less a deduction of £4900 which they claimed to be entitled to deduct as liquidated damages for the period from 20 May to 18 July, 1986. On 18 November 1988 the architect purported to give a new certificate that there was failure to complete the works by the completion date. This he clearly had no power to do because he was *functus officio* once he had issued the Final Certificate: *H. Fairweather Ltd* v. *Asden Securities Ltd* (1979). Before Judge Newey the defendants did not seek to rely on that certificate because they accepted that, once a final certificate had been issued, their architect had no authority to issue any further certificate.

Bell obtained summary judgment for the £4900. Judge Newey said that since the architect

> 'did not give a valid certificate of non-completion after fixing the completion date of 20 May 1986 and since the defendants were unable for want of a certificate to give a fresh notice, conditions precedent to their deduction of liquidated damages were not fulfilled and the plaintiffs are entitled to judgment.'

Liquidated damages but no loss

There is a growing body of case law about construction industry arbitrations, particularly as regards the JCT forms, and such decisions often cover points of great practical interest. One of the latest cases is *BFI Group of Companies Ltd* v. *DCB Integration Systems Ltd* (1987), which involved a challenge to an arbitrator's award in a dispute arising under a JCT Minor Works Form, and which emphasises that liquidated damages are recoverable for late completion even if the damage in the event is nothing at all.

The contractor (DCB) altered and refurbished offices and transport workshops for BFI. DCB claimed additional payment for the work and BFI counter-claimed in respect of delay and defective work. The dispute was referred to arbitration and, among other things, the arbitrator found that a ground slab had been defectively installed. In his award, he gave DCB the option of re-entering the premises and relaying the defective slab, as an alternative to paying damages for the cost of the remedial works and the consequential loss suffered by BFI.

He also held that DCB was in delay in completion but refused to award liquidated damages (as provided in the contract) on the ground that the employer had in fact suffered no loss as a result of the delay. This surprising conclusion was naturally challenged by the employer, who challenged the arbitrator's award on this and other grounds. BFI applied for leave to appeal under s.1(2) of the Arbitration Act 1979 on questions of law as to whether the arbitrator (who was a quantity surveyor) was right to refuse to award liquidated damages and also whether he was correct about an unrelated issue relating to payment for extra foundation works. The parties agreed that the application for leave to appeal and the actual appeal should be heard together, and this was duly done.

The liquidated damages issue is one which is commonly raised by contractors, and the very point about liquidated damages is that they are recoverable without proof of loss. However, contractors find it hard to understand why, if there is no loss, the employer should be entitled to recover any sum at all by way of liquidated damages even if they are in culpable delay, and clearly this is a view shared by some quantity surveyors and claims consultants. But it is not the law.

His Honour Judge John Davies QC, Official Referee, granted leave to appeal, although interestingly he emphasised that it would be an unsatisfactory approach to the problem of arbitration 'to be too pedantic about the legal reasoning of an arbitrator'. It was not for him to decide whether he would have done the same as the arbitrator or whether he thought that the arbitrator was wrong in his conclusion. The question was:

> 'Does it appear upon perusal of the award either that the arbitrator misdirected himself in law or that his decision was such that no reasonable arbitrator could reach?'

This was the test proposed by Lord Diplock in the well-known case of *The Nema* (1982) and, while the judge could not answer that question affirmatively on the extra foundations issue, he did so unhesitatingly on the liquidated damages point.

Clause 2.3 of the JCT Minor Works Form used provided that liquidated damages were to be paid if completion was delayed beyond the contract completion date or the extended date. The arbitrator found that the contractor was in delay because, although BFI was given possession of the building as a whole on the extended date for completion, two of six vehicle bays could not be used by the employer for another six weeks because necessary roller shutters had not been installed. Because BFI had to fit out the building after being given possession, the arbitrator ruled that it had not been caused any loss of revenue by the contractor's default and so was not entitled to liquidated damages.

The contractor's argument was that a liquidated damages provision did in fact presuppose some loss and that the only effect of a liquidated and ascertained damages clause was to quantify the loss in advance, provided that a loss could be shown in the first place. And, since the arbitrator found that there was no loss, liquidated damages did not come into play.

The learned judge rejected this argument. The object of a liquidated damages clause is to fix an amount which is a genuine pre-estimate of the amount of damage which the employer *might* suffer in certain circumstances. If those circumstances - late completion - occurred, it was irrelevant to consider whether there was in fact any loss. The liquidated damages provision was automatic - it was specifically designed to exclude any dispute about the loss.

Judge Davies said that he instinctively disliked liquidated damages provisions, but it was up to the parties concerned to decide whether they wanted such a clause.

If they included a liquidated and ascertained damages provision in the contract it came into play automatically if there was late completion. There was no question of the provision being a penalty and so unenforceable. The contract was one for the erection of six transport bays and two of them were not ready until six weeks after the extended completion date. The damages were pre-estimated in the contract and the employer was entitled to recover. The arbitrator was wrong in law in refusing any recovery of liquidated damages and so that part of the award was set aside.

Similarly, the learned judge ruled that the arbitrator had no jurisdiction to give DCB the option of re-entering the premises and re-laying the slab instead of paying damages. Quite apart from the fact that the order was impractical to implement, what the arbitrator had wrongly done was to substitute a new and different contractual obligation. The proper remedy was damages. The arbitrator had no power to order that if BFI did not allow the work to be carried out DCB should be relieved of its obligation to pay damages for the defective work.

This sensible and practical judgment will, it is to be hoped, finally still the controversy about liquidated damages being recoverable even if there is no actual loss. To my knowledge, it is the first reported construction industry case on the topic, but it has been the law for several centuries. A provision for liquidated damages is in fact advantageous to both parties; the employer does not have to prove any loss and the contractor knows the price he must pay for the breach of late completion.

Liquidated damages under JCT 80

JCT 80, clause 24.2, is a typical liquidated damages clause. It provides that, subject to the issue by the architect of a certificate of non-completion,

> 'the contractor shall, as the employer may require in writing . . . pay or allow to the employer the whole or such part as may be specified in writing by the employer of a sum calculated at the rate stated in the Appendix as liquidated and ascertained damages for the period between the completion date and the date of practical completion . . .'

Both the architect's certificate of delay and the employer's notice to the contractor are conditions precedent: *A Bell & Son (Paddington) Ltd* v. *CBF Residential Care & Housing Association* (1989).

An agreed damages clause of this sort is valid provided the sum is a genuine pre-estimate of the loss which will be caused to the employer by late completion or a lower sum, as is usually the case in construction industry contracts. Where a lower sum is specified – and often the amount is purely nominal compared with the potential loss – the clause also operates as a limit on the contractor's liability. This is a point which is frequently overlooked.

A liquidated damages clause thus defines the risk which the contractor is undertaking and also saves the employer the trouble and expense of proving his actual loss in litigation. It does not matter that the actual loss is hard to assess – and this assessment fails to be made at or before the time the contract is entered into. If the clause is valid, it is equally irrelevant that the employer's actual loss is greater or less or even if he suffers no loss at all. The genuineness of the estimate is judged at the time the contract is made and not at the date of the breach.

There is surprisingly little case law on the distinction between liquidated damages provisions and penalty clauses. A penalty clause is unenforceable. An agreed damages clause will be a penalty and not a genuine pre-estimate of loss:

(a) if the agreed sum is extravagant in relation to the greatest possible loss that could be suffered; or,
(b) as a general rule, if a single sum is payable in relation to various breaches with widely differing consequences.

This second principle is seldom of application in the construction industry because agreed damages clauses are normally limited to the breach of late completion.

There are problems in assessing the correct figure to insert as liquidated damages in the usual Appendix entry, particularly in the case of non-profit making projects such as educational buildings. In these cases it is thought to be acceptable to base liquidated damages on increased supervision costs, together with a return on the actual or notional capital employed, and various formulae are used by local authorities and others as a basis. Contrary to popular opinion, most liquidated damages figures are exceptionally low!

The contractor's liability to pay liquidated damages is, of course, modified by the operation of any extension of time clause. Thus, clause 25 of JCT 80 modifies both the contractor's liability to complete the works by the specified date (clause 23) and to pay liquidated damages for non-completion. Under JCT 80 too, the architect's certificate of non-completion under clause 24.1 is a condition precedent to the deduction of liquidated damages, which is also conditioned on written notice from the employer as provided in clause 24.2.1, as already noted.

The contractor will also be released from his obligation to pay or allow liquidated damages where the employer – through his architect – fails properly to exercise the power to extend time where the whole or part of the delay is in law attributable to the employer's fault, e.g., late receipt of instructions. The same principle would apply where there was no contractual power to extend time.

The case usually quoted in this connection is *Peak Construction (Liverpool) Ltd* v. *McKinney Foundations Ltd* (1970), where the architect was empowered to grant an extension of time if work was delayed 'in consequence of . . . other unavoidable circumstances'. Defective piling works caused work to be delayed, and the employer was in substantial delay in implementing its own experts' report. The Court of Appeal held that since the delay was partly caused by the employer, no liquidated damages were payable.

Lord Justice Salmon pointed out that a

> 'liquidated damages clause contemplates a failure to complete on time due to the fault of the contractor . . . If the failure to complete on time is due to the fault of both the employer and the contractor, in my view, the clause does not bite.'

He went on, and this is germane to JCT 80:

> 'If the extension of time clause provided for a postponement of the completion date on account of delay caused by some fault or breach

on the part of the employer, the position would be different. This would mean that the parties had intended that the employer could recover liquidated damages notwithstanding that he was partly to blame . . . In such a case the architect would extend the date for completion and the contractor would then be liable to pay liquidated damages for delay as from the extended completion date.'

That is, indeed, the usual position. But where no extension of time is granted when it ought to have been and where there is no extension of time clause, then the effect of delay caused by the employer is to release the contractor from liquidated damages and time becomes 'at large': see *Miller* v. *LCC* (1934). In that case the obligation would be to complete within a reasonable time, measured by reference to the disruptive effect of the delay: see *Dodd* v. *Churton* (1897). But if the liquidated damages clause becomes so inoperative, the common law right to unliquidated damages for breach is resurrected - though probably subject to a ceiling on the amount recoverable equal to the amount of the failed liquidated damages provision.

There must be a definite date from which liquidated damages are to run. If there is no date fixed by the contract, or time has become at large and the date fixed by the contract has ceased to be operative, the employer's right to recover the stipulated sum as liquidated damages has gone.

In fact, the proper operation of the usual extension of time clauses does cut down the contractor's potential liability to liquidated damages, but arbitrators and the courts are often asked to decide nice points on the topic. If, unusually, no figure for liquidated damages is inserted in the Appendix, the employer has to sue at common law and recover the actual loss he sustains on proof. A liquidated damages clause is helpful to both parties in defining the area of risk.

Chapter 6

Aspects of Prolongation and Disruption Claims

'Global' claims and awards

There is no doubt that a 'global' claim is permissible under both JCT and ICE contracts since the concept was approved by Mr Justice Donaldson in *Crosby* v. *Portland Urban District Council* (1967) and endorsed by Mr Justice Vinelott in *London Borough of Merton* v. *Stanley Hugh Leach Ltd* (1985). Both claims for time and money may be assessed on a global basis but this approach is not of universal application. JCT 80, clause 26, clearly permits the contractor to recover direct loss and/or expense in respect of an alleged disruptive event when it is not possible for him to tie in the various items of loss and/or expense the exact amount to be attributed to it.

However, a 'global' or 'rolled up' claim is only permissible if the contractor is able to satisfy the requirements identified in *Crosby*. Applied to JCT terms, Mr Justice Vinelott said that it was implicit

> 'in the reasoning of Mr Justice Donaldson that a rolled up award can only be made in the case where the loss or expense attributable to each head of claim cannot in reality be separated and secondly that a rolled up award can only be made where, apart from that practical impossibility, the conditions which have to be satisfied before an award can be made have been satisfied in relation to each head of claim.'

So, for example, under JCT 80 the contractor will need to have made a written application to the architect in respect of each event which he alleges gives rise to a monetary claim. A global award is not, in any case, allowable unless there has been an extremely complex interaction between the consequences of the various causative events so that it is difficult if not impossible to make an accurate apportionment between them.

The reason for using a global award was put thus by Mr Justice Donaldson:

> 'Extra costs are a factor common to all these clauses, and so long as the arbitrator does not make any award which contains a profit element [where this is not permissible] and provided he ensures that there is no duplication, I can see no reason why he should not recognise the realities of the situation and make individual awards in respect of those parts of the claim which can be dealt with in isolation and a supplementary award in respect of the remainder of these claims as a composite whole.'

In the preceding paragraph of his judgment he referred to the 'extremely complex interaction between the consequences of the various [events making it] difficult or even impossible to make an accurate apportionment of the total extra cost between the several causative events' and this is a *sine qua non* to a global settlement.

Architects and engineers can adopt the same approach in ascertaining or otherwise settling a contractor's claims. However, it seems to me that the contractor must in fact satisfy six conditions before a global approach can be adopted. This was obviously argued before the arbitrator in *Merton* v. *Leach* since there is reference to the point in the judgment.

Those conditions can be extracted from the cases, and in particular from *Crosby*. They are:

- The actual basis of each claim must be defined separately. This was clearly done in *Crosby* where Mr Justice Donaldson referred to the paragraphs in the contractor's points of claim in the arbitration which defined the various matters giving rise to the claim as well as to the arbitrator's findings of fact.
- Notice must have been given by the contractor in respect of each claim, whether it be for time or money or both. In *Crosby*, again, the arbitrator had found that the requisite written notice had been given to the engineer in respect of all the claims but one and Mr Justice Donaldson held that the lack of notice was fatal to the claim.
- The items of loss and expense must be such as are difficult if not impossible to quantify. Items which are reimbursable elsewhere under the contract must be excluded. This test was again satisfied in the *Crosby* case where the arbitrator had made awards on the quantum involved and had made due allowance for those items which had been valued elsewhere under the contract.
- The loss and expense must relate to specific and precise periods of delay or disruption.
- It must be either difficult, or impossible, to make an accurate

apportionment of the total extra cost between the several causative events.

- If possible, individual awards should be made for those parts of the claim which can be dealt with in isolation and only the remaining items can be covered by a global or composite award.

These conditions must be satisfied before a global approach can be used. If they are met, if application is made for reimbursement of direct loss and/or expense attributable to several heads of claim and it is impracticable to disentangle the part directly attributable to each head of claim, the global loss must be ascertained and paid so that the contractor is not unfairly denied his due.

However, as the learned editors of *Building Law Reports* point out in their commentary on *Crosby*:

> 'It is very tempting to take the easy course and lump all the delaying events together in order to justify the total overrun or total financial shortfall. *That argument is justifiable only if the alternative course is shown to be impracticable.*' (Emphasis supplied.)

Applying for information timeously

Contractors' claims under JCT contracts inevitably give rise to controversy for, despite their widespread usage, the wording of the money claims clause is ambiguous in many respects. Indeed, this stricture applies equally to the majority of clauses in all standard form contracts.

Case law guidance is, therefore, welcome as so many important points are settled privately in arbitration. *London Borough of Merton* v. *Stanley Hugh Leach Ltd* (1985) is a case which casts light into dark corners, and gives guidance on many points of interest. It was an appeal from an arbitrator who had to decide no less than 14 preliminary legal issues. Merton appealed against his interim award.

The contract was in JCT 63 form, somewhat amended, and was for the erection of 287 new dwellings at Mitcham. The contract sum was £2,265,217 and the contract period as agreed was effectively 33 months. The date for possession was 15 September 1972 giving a contract completion date of 14 June 1975. Practical completion occurred on 20 May 1977, 101 weeks after the due date.

There were disputes about extensions of time and the usual allegations and cross allegations giving rise to various preliminary issues of law. The first of these related to the status of Leach's programme for completion of the whole works. Opposite each activity were a number of conventional signs indicating the dates on which drawings and other information would be required. The question before the court was

> 'Did [the programme] become a specific application for instructions, drawings, details or levels within the meaning of clause 23(f) and 24(1)(a) and, if so, when?'

The main point at issue, of course, was whether it was a specific application made

> 'on a date which . . . was neither unreasonably distant from nor unreasonably close to the date on which it was necessary to receive the same'.

The judge, agreeing with the arbitrator, saw no reason why a document in diagrammatic form, should not be treated as a 'specific application'. What the provisions call for, he remarked,

> 'is a document which indicates whether by words or by the use of conventional signs, or in any other form, what the contractor requires and when he requires it, and which does so in sufficient detail to enable the architect to understand clearly what is required of him.'

But was the application made at the right time? It was submitted very early in the contract and related to the whole contract period. As the judge noted, the architect is not required to furnish information unreasonably far in advance of the date when the contractor requires it in order to carry out the work efficiently. But the wording of the clause is to ensure that the architect is not troubled with applications too far in advance of the time when the information is actually needed by the contractor, and to ensure that he is not left with insufficient time to prepare them.

On this basis, said Mr Justice Vinelott,

> 'there seems to me to be no reason why an application should not be made at the commencement of the work for all the instructions etc., which the contractor can foresee will be required in the course of the works *provided that the date specified for delivery meets these two requirements.*'

The emphasis is mine for, as the judge later remarked, if

> 'the works do not progress strictly in accordance with his plan, some modification may be required to the prescribed time-table, and the subsequent furnishing of instructions and the like.'

The case is not, therefore, authority for the view that in every case a blanket application made early in the contract meets the requirements of the clauses. The arbitrator had held that it was a specific application only in respect of 'earlier applications', and the judge expressly cautioned that it did not follow that the programme was 'a sufficiently specific application in relation to every item of information required', more particularly in the light of the delays and the re-arrangement of the programme of work.

Programmes cannot be considered in isolation. They are not contract documents, but the judge said that the programme should be considered in conjunction with other communications by Leach. The question was whether programme and letters, etc., together

> 'constituted a sufficiently specific application in relation to each item of information there specified, and whether it was in the light of the history of the carrying out of the works requested at a time which met the [contract] requirements.'

Contractors should not, therefore, be misled into thinking that a blanket application is necessarily or even possibly sufficient. The application must be specific and it must be made at the correct time. The application must not be too early, or too late, and *London Borough of Merton* v. *Leach Ltd* certainly does not support the argument that a blanket application at contract commencement, or the giving of an information supply programme or chart to the architect is sufficient.

The timing of the application is important because the architect's work must not be unnecessarily disrupted. In some cases it may be reasonable to require finishing details on day 1, but an application which is too far in advance may reasonably be rejected. Certainly, contractors may (and should) use their programmes to indicate dates on which information is required from the architect. This will be a 'specific application'. The time constraints are just as important, however.

It is always difficult to know where to draw the dividing line about the timing of applications. It cannot be said that a request for all information early on is a 'continuing application'. The time constraints are to be determined having regard to the date for completion as originally fixed or extended, and must always be considered against the current completion date.

'As much as it's worth'?

Contractors' claims for delay and disruption are sometimes made on a *quantum meruit* basis, i.e., 'as much as it is worth'. The contractor claims the price which he would have charged for the job, had he known in advance the extra problems which he was likely to encounter.

The *quantum meruit* approach to a contractor's claim for damages for disruption and delay is deceptively simple – and it is very attractive from the contractor's point of view. Contractors using this basis of claim refer to the Canadian case of *Penvidic Contracting Co. Ltd* v. *International Nickel Co. Ltd* (1975), where the Supreme Court of Canada held that the contractor was entitled to damages on the basis of the higher initial estimate than would have been made had the likely difficulties he actually encountered been known. These difficulties were caused by the employer's breaches of contract and there was evidence before the court that the figure claimed was a reasonable estimate.

The contract was to lay ballast and track for a railroad, and the employer was in breach of his obligation to facilitate the work. The contractor had agreed to do the work for a certain sum per ton of ballast, and claimed by way of damages the difference between that sum and the larger sum that he would have asked had he foreseen the adverse conditions caused by the employer's breach of contract. There was evidence that the larger sum would have been a reasonable estimate.

Both at the trial and on appeal to the Supreme Court, it was held that where proof of the actual additional costs caused by the breach of contract was difficult, it was right to award damages on this basis. The difficulties of accurate assessment cannot relieve a wrongdoer of the duty of paying damages for breach of contract. All this really says is that in wholly exceptional circumstances it may be possible to adopt a 'broad-brush' approach to the assessment of individual heads of damage where proof of each and every head is very difficult, if not impossible.

The reasoning of the Supreme Court is interesting, the difficulty in fixing the amount of damages being well known. The basic principle is that the injured party 'is to be placed, as far as money can do it, in as good a situation as if the contract had been performed,' as Viscount Haldane LC put it in the House of Lords in *British Westinghouse Co. Ltd* v. *Underground Railways Co. Ltd* (1912).

The courts, both in England and Canada, have gone a long way in holding that difficulty in ascertaining the amount of the loss is no reason for not giving substantial damages, as is well shown by the

English case of *Chaplin* v. *Hicks* (1911), where the plaintiff accepted the defendant's offer to apply for selection for employment by him as an actress. The plaintiff was one of 50 applicants from whom 12 were to be selected. She was given insufficient notice of the interview at which the final choice was to be made.

The Court of Appeal held that she was entitled to £100 damages for breach of contract because the damage which she had suffered, i.e., loss of opportunity, was within the parties' contemplation. Although it was difficult, if not impossible, to arrive at an accurate assessment of the amount of damages, the question had rightly been left to the jury which (in those days) assessed damages.

The Canadian Supreme Court cited this decision with approval in the *Penvidic* case, and also cited *Wood* v. *Grand Valley Railway Co.* (1913), to the effect that where it is impossible to estimate with anything approaching mathematical accuracy the damages sustained, the judge (or arbitrator) must do 'the best it can' and his conclusion will not be set aside even if 'the amount is a matter of guess work'.

In *Penvidic* the Supreme Court ruled that there was no objection to using the contractor's method of assessing

> 'the damages in the form of additional compensation per ton rather than in attempting to reach it by ascertaining items of expenditure from records which, by the very nature of the contract, had to be fragmentary and probably mere estimations.'

This is no more than saying that a 'global' approach is permissible to a contractor's claim where it depends on 'an extremely complex interaction in the consequence of various denials, suspensions, and variations [so that it is] difficult or even impossible to make an accurate apportionment between the several causative events': (*J. Crosby & Sons Ltd* v. *Portland UDC* (1967), approved in *London Borough of Merton* v. *Stanley Hugh Leach Ltd* (1985).

The *Penvidic* case does not, therefore, support the *quantum meruit* basis of claim. As Emden's *Building Contracts*, 8th edition, Vol. 2, N/46A puts it:

> 'Legally the weakness of this approach [i.e., a *quantum meruit* one] lies in its according to the contractor a right, in effect, to profitable remuneration considerably in excess of that contracted for, whereas normally damages for breach of contract are purely compensatory of loss which has been suffered . . .'

Moreover, claims under JCT and ICE terms are rights given under the express terms of the contract. If the contractor is to succeed in such a claim he must comply with the procedural provisions of the contract.

Under JCT 80, for example, this means that the claimant must provide details of the 'direct loss and/or expense': see the duty expressly spelled out as regards final adjustment of the contract sum in clause 30.6.1.1. The contractor's obligation in a claim situation is to provide all the information which the architect or quantity surveyor reasonably requires for the purposes of his ascertainment.

Not all the 'claim' items under ICE and JCT terms are breaches of contract; the claims clauses confer rights to payment in respect of some items which are not breaches of contract at common law, e.g. clause 26.2.5. If the contractor opts to proceed under the contract, and thus gets the benefit of payment in interim certificates, then he must comply with the procedural provisions.

The *Penvidic* approach could be helpful if the contractor elects to proceed at common law in a few rare cases: see Hudson's *Building Contracts*, 10th edition, p. 601; '*In cases where the work is partly carried out and the contract is repudiated*' contractors are advised to consider whether to sue for breach or on a *quantum meruit* basis. Where

> 'the contract rates or price are low or uneconomic, it may well be that a reasonable price for work done will be more advantageous . . . particularly if a substantial amount of work has been done prior to the employer's repudiation.'

Repudiation is a different creature from a regulated contract claim; the contractor's *quantum meruit* approach should be rejected as a method of assessing such claims, and the limitations of the 'global approach' to JCT and ICE claims should be noted.

Programme problems

Most standard contracts provide the employer with little effective control over the contractor's rate of progress, but one sometimes finds a contractual requirement for the contractor to submit a programme to the architect or engineer for approval. This can be done under JCT forms; JCT 80 contains the optional provision for submission of a master programme, but the programme is not a contract document. Its status seems purely evidentiary as to intent and deviation from the programme is not a breach of contract and in practice it has very little effect in computing extensions of time or, more controversially, disruption claims.

In the well-known case of *Glenlion Construction Ltd* v. *The Guinness Trust*

(1987), a provision in the Bills required the contractor to submit a programme to the architect showing a completion date 'no later than the date for completion' and the High Court held, very sensibly, that, while the requirement was a term of the contract and the contractor was entitled to complete the works earlier than the contract completion date, there was no corresponding obligation on the employer (or his architect) to so perform the contract as to enable the contractor to do so. The contract completion date was the critical one. Under JCT terms at least *Glenlion* knocked on the head contractors' claims for 'acceleration of programme'.

The other side of the coin is illustrated by an interesting decision of His Honour Judge John Newey QC in *Kitson Sheet Metal Ltd* v. *Matthew Hall Mechanical & Electricial Engineers Ltd* (1989) which arose out of the building of Terminal 4 at Heathrow Airport. Matthew Hall were works contractors for the installation of pipes and ducts and had sub-contracted part of their work to Kitson.

In the negotiations leading up to the sub-contract, Matthew Hall wrote a letter to Kitson requiring 'all work to be carried out in accordance with our requirements and to suit the progress of the main contract works'. The first preliminary programme submitted reflected a similar document submitted by Hall to the management contractor and showing start and finish dates.

A subsequent pre-contract programme sent by Hall varied the dates and showed areas available for work. No contract was signed, but Kitson started work, despite the fact that areas were not available to them as provided by the programme. Six months after start, Matthew Hall sent what was described as 'an order' which made no reference to the programme but required the 'work to be carried out in accordance with the dictates of our management team, to enable the overall mechanical services to be complete and handed over to our client on 18 March 1985' which was later than the programmed completion date. Later Matthew Hall sent other programmes which differed materially from the earlier versions. Kitson had been working on the site all the time.

Nearly 17 months after Kitson had started work they entered into a written agreement with Matthew Hall which contained explicit terms as to completion. Hall was to provide the necessary access and work areas and Kitson undertook to 'commence the execution of the works . . . and . . . thereafter proceed . . . with due diligence and without delay, except as may be expressly sanctioned or ordered by [Hall] or be wholly beyond the control of [Kitson] . . . [and] shall complete the works . . . as detailed in [Hall's] order . . .'. Kitson did not complete the work until 18 March 1985.

Kitson sued Hall claiming payment for measured work, variations under the contract – the variations were alleged to have been worth £1.1 million – and breach of contract. The variations clause defined the term 'variation' in more or less standard JCT terminology as including 'the alteration of the manner or sequence of working' and whether there were orders for such variations was a major issue before the court. As Judge Newey put it:

> 'in essence the issue raises the question of whether [Kitson] were entitled under the contract to work to a programme and whether any written order requiring the departure from it constituted a variation.'

The learned judge ruled that Kitson were not entitled to work to any programme. The programme was not a contractual document. Matthew Hall could give Kitson instructions as required and their doing so would not constitute a variation based on a change in the manner or sequence of the work. Kitson were contractually bound to work to Matthew Hall's instructions and their claim for payment for variations was rejected.

It may well be said that dispute was almost inevitable having regard to the contractual background, but lengthy negotiations, changes of mind and requirements and so on are not unusual in 'one-off' contract situations. The link between the contract and actual construction management remains very weak and in most situations at the very best any programme requirement can only be of limited value.

Under JCT contracts provisions in bills with respect to programmes, even where there is a programming requirement or phased handovers within the main contract completion period will not normally give rise to any time-related obligation and at the very best in a claims situation the programme is only going to be of use for general monitoring purposes. In my experience under JCT 80, neither the original nor amended master programme is usually sufficient to substantiate a monetary claim arising from any particular cause of delay or disruption. It is merely of *some* evidentiary value.

Late information

Under any standard construction contract it is an implied term that the architect (or engineer) will issue to the contractor in due time such instructions, drawings, details and other information as the contractor needs to carry out the works in accordance with his obligations under the contract. Breach of this implied term will entitle the contractor to damages at common law.

In JCT and ICE contracts provision exists for the contractor who has been delayed by late information to be granted extensions of time and, separately, to be paid for any direct loss he suffers as a direct result. In JCT 80, of course, his contractual right to reimbursement is subject to a requirement that he should have made specific written application for the information at the right time: see *London Borough of Merton* v. *Stanley Hugh Leach Ltd* (1985). The contract acknowledges and makes provision for the possibility that drawings, etc., will be issued by the architect at times other than those which will enable him to comply with his contractual obligations.

JCT 80, clause 5.4 sets out the time when the architect is to furnish further information, i.e. 'as and when from time to time may be necessary', while ICE, 5th edition, refers to the issue of information by the engineer 'at a time reasonable in all the circumstances'. Clearly, these words enable the architect or engineer to issue this further information piecemeal and from time to time, i.e., during the currency of the contract, and in practical terms this means that in JCT contracts an implied term as to supply of further information is unnecessary. It would merely be repetitive of the express obligation in the contract requiring the architect 'as and when from time to time may be necessary [to] provide [the contractor with] further drawings or information as are reasonable necessary . . . to enable the contractor to carry out and complete the works in accordance with the Conditions'. Breach of this express term may give rise to a claim for damages at common law and so the contractor does not need to rely on the contractual provisions for reimbursement. Furthermore, in order to found a common law claim the contractor need not have applied for the information in writing 'on a date which having regard to the completion date was neither unreasonably distant from nor unreasonably close to the date on which it was necessary to receive the same' (JCT 80, clause 26.2.1 - reimbursement of direct loss and/or expense; clause 25.4.6 - extension of time). Clearly, too, if the architect is late in supplying necessary information he is bound to grant an appropriate extension of time even if the contractor has given no notice of claim under clause 25.2.1 since his (the architect's) default is employer responsibility.

Under JCT terms this leaves for consideration the phrase 'as and when . . . may be necessary'. In *Neodox Ltd* v. *Swinton & Pendlebury UDC* (1958) - a civil engineering case - Mr Justice Diplock held that the convenience of the engineer and the order in which he determined the works should be carried out were factors to be taken into account in determining what was a reasonable time. Under JCT contracts the architect does not control the order of working and thus the word 'necessary' must primarily be interpreted from the point of view of the contractor. In the old case of *Wells* v. *Army & Navy Co-operative Society* (1902), the contract did not actually prescribe the times when detailed drawings were to be provided to the contractor by the architect and, on appeal, Lord Justice Vaughan Williams rejected the architect's view that progress on site was a controlling factor in the timing of the supply of drawings; failure to provide plans and drawings in due time was held to be a breach. Mr Justice Wright held that the contractors were entitled to receive the drawings 'promptly upon request . . .' and this has led some people to argue that a request is also necessary. This view is incorrect.

In considering the right timing from the contractor's point of view, it is quite clear that 'necessary' is a much stronger word than 'reasonable', which infers some degree of balancing exercise as that envisaged in the *Neodox* case where the matter of timely supply had to be considered from the point of view of both the engineer and his staff and that of the contractor.

Under JCT 80 (and IFC 84) terms, I am firmly of the view that regard must be had to, *inter alia*, three matters before the contractor's actual progress on site needs to be taken into account. These are:

- His obligation to complete the works in accordance with the contract.
- The time necessary for him to organise adequate supplies of materials, labour and plant.
- The carrying out of any necessary prefabrication or preparation of materials.

The last two factors must be considered having regard to the time necessary to ensure that the items are available for incorporation into the works so as to enable him to fulfil his overall contractual obligation. *Merton* v. *Leach* is authority for the view that clause 5(4) of JCT 80 imposes on the architect an obligation not only to furnish the contractor with information as and when necessary, but also that the architect will act with reasonable diligence and will use reasonable skill and care in providing the information, and that the printed contract does not contain a complete code of his obligations.

Access and claim denied

JCT 80, clause 25.4.12, which enables an extension of time to be granted to the contractor as a result of the employer's failure 'to give in due time ingress to or egress from the site of the works' in specific and clearly defined circumstances.

This has nothing to do with failure to give possession of the site, but is concerned merely with access to it. The contractor may also claim loss and expense under the corresponding paragraph in clause 26.

The land, etc., must be in the employer's possession *and* control as well as being adjoining or connected to the site. Clause 25.4.12 does not apply, therefore, if the employer fails to obtain a wayleave over land not in his possession, nor does it deal with the situation where the bills or drawings do not lay down the contractor's rights and obligations as regards access.

It seems, therefore, that in such a case the position would have to be dealt with at common law: failure to give access might amount to failure to give possession of the site.

The case of *LRE Engineering Services Ltd* v. *Otto Simon Carves Ltd* (1981), arose from a special form of contract, clause 24 of which provided that 'access to and possession of the site shall be afforded to [LRE by OSC] in proper time for the execution of the work. [LRE] shall have access by means only of those roadways and railway sidings of [OSC's client]'. The clause further provided that access and possession 'shall not be exclusive to [LRE] but only such as shall enable the contractor to execute the work'.

LRE was carrying out work at the Port Talbot Steelworks and it had completed 90% of it under its sub-contract when the 1980 steel strike occurred. This lasted (with a short intermission) from 3 January to 8 April. As a result of the activities of the pickets on the site, LRE was prevented from completing its work. When the strike ended, LRE had not only to complete the work but it had to put it back in the condition it was in at the date when the strike broke out. The contractor estimated these extra costs to be £320,000.

LRE alleged that by being denied access to the site because of the activities of the pickets, OSC was in breach of contract. OSC countered this argument by saying that first, it was not in breach of contract as alleged, and second that LRE was itself in breach by repudiating the contract. A very distinguished arbitrator found against LRE, and his view was upheld by Mr Justice Goff in the High Court. There was no breach of contract by OSC because the access was impeded by third parties – the pickets.

Mr Justice Goff, having reviewed the arbitrator's findings of fact, expressed this view of the meaning of clause 24:

> 'It appears to me that . . . first of all there was a physical means of access – in other words there were roadways reaching from the

perimeter of the Port Talbot Works to the site where the contract work was to be carried out; and it appears also . . . that OSC did provide the contractor, LRE, with the opportunity of entering the site by such means of access. What did happen, however, was that the activities of a third party prevented LRE from taking advantage of the opportunity that was made available to them by OSC . . . [The fact] that they were so prevented does not, it seems to me, amount to any breach of the obligation undertaken by OSC . . .'

This is the crux of the judgment but another very interesting point was canvassed, namely had there been a breach by OSC, would it have entitled LRE to put the contract to an end? The arbitrator had found that even if a significant number of LRE's workmen had crossed the picket lines, it would not have been able to work because BSC's security staff would have withdrawn their labour in sympathy with the strikers, thereby rendering further work impossible.

OSC said that it followed from this that even if there was a breach of clause 24 on its part, nevertheless LRE could have done nothing had it got into the site. In those circumstances the breach, if there was one, could have caused no damage. The breach was not 'fundamental' and would not have entitled LRE to bring the contract to an end. 'That submission is, in my judgment, well founded,' said Mr Justice Goff.

The principles discussed in *LRE Engineering Services Ltd* v. *Otto Simon Carves Ltd* apply equally to JCT 80. The contractor is not entitled to wholly unrestricted access, but only to the extent laid down in the Bills and drawings or as agreed. The access must have been prevented by the employer or by someone for whom he is vacariously responsible in law.

Similar problems may well be arising elsewhere. The same principles would apply to the question of possession of the site as opposed to access to it.

This is well-illustrated by the old case of *Porter* v. *Tottenham Urban Council* (1915). In that case, Porter agreed to build a school for the council on land belonging to the authority. The contract provided that he should be entitled to enter on the site immediately and he was to complete the work by a specified date. The only access to the site was from a road, and as the soil of the council's land adjoining the site was soft, the contract stipulated that Porter was to lay a temporary sleeper roadway to the site.

Porter began work but was forced to abandon it because of a threatened injunction from an adjoining owner, who claimed that the site of the road was his property. The third party's claims were held to be unfounded. Porter then resumed and completed the work, and later claimed damages against the council in respect of the delay caused by the third party's action.

The Court of Appeal held that the claim must fail. There was no implied warranty by the council against wrongful interference by third parties with free access to the site. The employer was not in breach of the obligation to give sufficient possession to the contractor where, as here, the contractor was wrongfully excluded from the site by third parties for whom he is not responsible in law and over whom he has no control.

The position is different, of course, if the employer has control of the site and can take effective steps, as for example, where squatters occupy the site and so prevent the employer giving possession: *Rapid Building Group Ltd* v. *Ealing Family Housing Association Ltd* (1984).

As in so many other cases, it is all a question of the allocation of risks: in many instances though there may be loss, there is no recovery.

Claims against sub-contractors

Delegates' questions at seminars always highlight the widespread use of non-standard forms of sub-contract, usually devised by main contractors for their own advantage. The inclusion of 'pay when paid' clauses is a notorious example and, rightly or wrongly, such clauses are violently objected to by sub-contractors. Set-off clauses allowing the main contractor to deduct 'claims' allegedly due to him are another example.

From the lawyer's point of view, of course, the main thing is to ensure that the terms of the main contract are consistent with those of the sub-contract. This is best achieved by using the industry-wide standard forms designed to be used with one another – forms which have the merit of being agreed by representatives of all users. Sub-contractors must remember that the legal relations between them and the main contractor will be governed by the sub-contract terms and not by terms which may have been intended to be incorporated. The main contract provisions are not enforceable by the sub-contractor.

Difficulties can arise even where the industry's standard forms are used, particularly in the area of sub-contract claims in respect of delay and disruption. Where the JCT family of forms is used, the principles governing loss and expense claims are the same under both main and sub-contracts. The only differences are procedural, and it is for the parties to ensure that they observe the special procedural requirements of the particular form.

The best example is provided by looking at the terms of the JCT nominated sub-contract standard form NSC/4, clause 13.1 of which deals with claims by the sub-contractor which must be passed up the contractual chain to the employer by the main contractor. It mirrors the main contract provision in clause 26 of JCT 80; and in principle at least no problems should arise. (I know that they do in practice.) Real difficulties arise under clauses 13.2 and 13.3, which cover sub-contractors' claims against the main contractor and vice versa.

Clause 13.3 – covering main contractor's claims – is often a source of friction and is widely misunderstood. As usual, if refers to disturbance of regular progress of the works caused by 'any act, omission or default' of the sub-contractor or those for whom he is responsible in law. A major difficulty is that it envisages the claims being *agreed* and it is only when they are so agreed that the amount 'may be deducted from any monies due or to become due to the sub-contractor or may be recoverable from (him) as a debt'.

A common area of misunderstanding is that many risks must be borne by the main contractor himself; clause 13.3 enables the main contractor to claim from the sub-contractor for direct loss and/or expense where the regular progress of the works is materially affected by the sub-contractor's default, and contrary to a commonly-held view this does not mean that the sub-contractor becomes liable for liquidated and ascertained damages under the main contract.

The main contractor's right to claim for delayed completion of the *sub-contract works* is dealt with by clause 12.1 and it is dealt with on a different basis from his right to claim for the effect of default by the sub-contractor upon regular progress of the *main contract works* under clause 13.3.

Indeed, JCT 80, clause 25.4.7, expressly provides that 'delay on the part of nominated sub-contractors or nominated suppliers' is a relevant event for an extension of time. If the main contractor applies for and is granted an extension of time under the main contract he suffers no liability to liquidated damages for the nominated sub-contractor's delay and hence there are no liquidated damages to pass on. And main contractors must read clause 25.4.7 carefully: an extension of time is only allowable in respect of nominated sub-contractors' delays which they have taken all practicable steps to avoid or reduce.

Delay by the nominated sub-contractor may, indeed, be an act or default by him, but this does not mean that the sub-contractor picks up the tab for liquidated damages under clause 13.3. There are no liquidated damages to pass on.

The main contractor's right to claim for delayed completion of the sub-contract works under clause 12 depends on the issue by the architect of his certificate under main contract clause 25.15.1, a point often overlooked by over-enthusiastic main contractors. Their answer is to read the contract.

Moreover, as John Sims and I have pointed out (*Building Contract Claims*, 2nd edition p. 270):

> '. . . The architect is only to issue the certificate if he is satisfied that the provisions . . . regarding extensions of time have been properly applied, and the architect is therefore given a period of two months before he may become obliged to issue the certificate. The architect is thus clearly placed under an obligation to make proper investigation to ensure that the main contractor has passed to him all notices and applications from the sub-contractor (including those alleging delay caused by the main contractor's own default) and that all extensions of time for which he has given permission have been properly granted'.

In those cases where the main contractor does have a valid claim against the sub-contractor under clause 13.3, and is unable to get the sub-contractor's agreement, the sub-contract provides its own remedy. Clause 23.2 covers those cases where no agreement is reached and the main contractor has quantified its amount 'in detail and with reasonable accuracy'. Then, at least 20 days before the payment from which he proposes to set-off the claim, he gives written notice to the sub-contractor, and it must be of loss and/or expense which he has *actually* incurred.

If the main contractor can bring himself within the provisions of clause 23.2, gives his notice and the sub-contractor disagrees, then the matter goes to the adjudicator under clause 24 – provided the sub-contractor complies with the very complex provisions of that clause. These procedures – with valuable protection for both disputants – are, of course, seldom found in non-standard forms.

The right to a reasonable sum

There is no implication that the costs incurred by the contractor in preparing his tender will be reimbursed by the employer, but the contractor may be entitled to reasonable payment if, at the employer's request, he does substantial preparatory work over and above normal tender preparation and no contract is ever placed.

This elementary principle follows from *William Lacey (Hounslow) Ltd* v. *Davis* (1957), where a contractor tendered for rebuilding war-damaged premises. His tender was not accepted but at the defendant's request he prepared various schedules and estimates which the employer used in negotiations with the War Damage Commission. The defendant then placed the contract elsewhere. Mr Justice Barry held that the plaintiff was entitled to payment of a reasonable sum (£250) for the work he had done subsequent to tender.

A greater amount was at stake in *Marston Construction Co. Ltd* v. *Kigass Ltd* (1989), where the plaintiff contractors claimed £25 946.23 for preparatory works over and above the costs and work of preparing their tender for a design and build contract to provide a factory at Warwick to replace one which had burnt to the ground.

The contract for the work was never placed because the proceeds of the insurance money from the old factory were insufficient to cover the cost of rebuilding to the local authority's standards.

The plaintiff's tender was the best value for money but because there was a tight timescale it needed to be supplemented with further details. The tender was followed by a vital meeting on 18 December 1986, when the plaintiff was the only tenderer invited to discuss its tender. The defendant's chairman made it clear that no contract to rebuild the factory would be entered into unless and until he had obtained the insurance money to pay for the rebuilding. Nothing was said which indicated that in the event of the insurance money not being forthcoming all preparatory work done by the plaintiff up to that point would be at the contractor's risk.

The defendant was aware the plaintiff would have to start preparatory work before the contract was signed, and it was understood if not spelt out that there should be a four-week lead-in period.

It was expressly agreed that the contract would be signed (if the insurance money had been received) on 16 January 1987 and that work would start on site on 2 February 1987. It was also found that it was made plain to the plaintiff at the meeting that, subject to payment of the insurance money and the employer's surveyor being satisfied on points of detail normal to a tender appraisal, the contract would be given to the plaintiff.

Judge Peter Bowsher QC found in favour of the contractor. There was an express request by the defendant to the plaintiff to carry out a small quantity of design works and there was an implied request to carry out prepartory works in general. Both requests gave rise to a right to payment of a reasonable sum, the amount to be assessed subsequently.

The learned judge said that both the plaintiff and the defendant believed that the various conditions would be satisfied and, in the words of Mr Justice Barry in the *Lacey* case:

> '... the proper inference from the facts proved in this case is not that this work [i.e. the preparatory work] was done in the hope that this building might possibly be reconstructed and that the plaintiff company might obtain the contract, but that it was done under a mutual belief and understanding that this building was being reconstructed and that the plaintiff company was obtaining the contract'.

The plaintiff's claim was in restitution and not in contract; the defendant obtained the benefit of the plaintiff's preparatory work and in His Honour's judgment the fact that it was merely a potential benefit was not relevant. After considering the conflicting academic views he said:

> 'It seems to me that in appropriate cases, the benefit may consist in a service which gives a realisable and not necessarily realised gain to the defendant particularly when, as here, the service is part of what was impliedly requested'.

The preparatory works were done for the defendant's benefit and were only done for the contractors' benefit in the sense that they hoped to make a profit out of them. Because of the works, progress was made towards getting consents and in the end the defendant's surveyor had in his hands some designs and working drawings together with an implied licence to build to those drawings. 'Whether the defendant decides ultimately to build a factory or to sell the land, it has a benefit which is realisable.'

The facts of the case, although different in important respects, were similar in kind to those in *Lacey's* case. There was a request to do the work, though the request in respect of the bulk was implied rather than express. It was contemplated that the costs of the preparatory work would be paid for out of the eventual contract.

Both parties believed that a contract was about to be made despite the fact that there was a clear condition which had to be met by a third party (the insurer) if the contract was made. But the defendant obtained the benefit of the work and so the contractor was entitled to be paid.

Marston Ltd v. *Kigass Ltd* will take its place in the law books as a leading authority on a contractor's entitlement to payment of a reasonable sum for work done at the employer's request where there is no contract.

Chapter 7

Disputed Heads of Claim

Loss of productivity

Money claims under JCT 80, clause 26 are a controversial topic and employers sometimes allege that the philosophy of some contractors is 'tender low, claim high'. In fact, claims under clause 26 are closely circumscribed. The contractor must comply with the procedural mechanism of the clause and he must establish that he has suffered or incurred '*direct* loss and/or expense' as a result of the event relied on. Clause 26 confers a *right* to extra payment provided its mechanism is observed.

One of the areas of controversy results from the fact that loss and expense claims under JCT forms have traditionally been linked to extensions of time and this erroneous view is now compounded by clause 26.3, which has the sidenote 'relevance of certain extensions of completion date'. In truth, there is no connection between clause 26 and the extension of time provision, clause 25. An extension of time under clause 25 does not entitle the contractor to any extra payment, whether by way of 'extension of preliminaries' or otherwise. Claims under clause 26 are quite separate and distinct.

Clause 26 speaks of 'regular progress of the works or any part thereof' being 'materially affected' by one or more of the events listed in clause 26.2, and some legal authorities argue that this must necessarily involve *delay* in progress, thus cutting out loss of productivity as a head of claim. This view is put forward by Mr I.N. Duncan Wallace in his *Building and Civil Engineering Standard Forms*, p.112, where he says that 'it is limited to loss or expense due to *delay to progress* and does not cover cases where the work, without progress being affected, has become more expensive or difficult'. He adds that this is unfair and that the phrase, no doubt, 'may be interpreted liberally in the contractor's interest, but it is clear that it does not cover loss of productivity'.

With respect, I do not find this line of argument convincing.

'Progress' means 'move forward or onward; be carried on', and it seems plain to me, and to others, that 'regular progress of the works' can be 'materially affected' without there being any delay in completion at all – the word 'delay' is not used in the clause. Mr Duncan Wallace himself provides a good example of the sort of thing that, in his view, does not give rise to a claim and contrasts it with the situation under JCT 63, clause 11(6), under which he allows that loss of productivity is a head of claim. (In both cases he is writing of the corresponding 1963 provision.)

A late delivery of a vital instruction may well cause the contractor to incur additional cost because he cannot use the labour and plant resources available to him in an efficient way. Late instructions amount to a breach of contract in any event, and in my view a claim under clause 26.2.1 is not limited to heads of damage caused by delay to progress: it covers other possible consequences of the late instruction, including loss of productivity. Claims under clause 26 can be equated with claims for damages at common law, and are governed by the same principles. On that basis, the consequences of the late instruction must have been within the reasonable contemplation of the contracting parties at the time the contract was made. The additional cost arises as a direct and natural result of the architect's default and is thus claimable under the contract terms: see *Hadley* v. *Baxendale* (1854).

The contractor must, of course, be able to *prove* loss of productivity, and this may be difficult to do. But assuming that he can provide the necessary evidence, I submit that loss of productivity is a valid head of claim. Clause 26 permits claims for both *disruption* and *delay* (or prolongation). A claim under the first head arises where the employer (through his architect or otherwise) makes work more difficult or expensive than foreseen; a prolongation claim arises where the work is actually delayed and the completion date is affected.

In practice, it is often difficult to separate the two types of claim because claims for disruption and delay often arise from the same fact. Indeed, the position may be further complicated by the presence of extra work, i.e., variation work.

How is the architect or, (more usually) the quantity surveyor to ascertain and assess the direct loss and/or expense in such circumstances? It is up to the contractor to provide him with the necessary information from records. However, there are very considerable difficulties in calculating the precise loss arising from both delay and disruption under this and, indeed, other heads. Emden's *Building Contracts and Practice*, 8th edition, Vol. 2., p.N/45 puts forward a method of calculation based on loss of productivity:

> 'Initially, a period is examined when the contract was running normally, and the value of work done during that period is assessed and then divided by the number of operatives and/or items of plant on site. The figure thus arrived at is compared with the same figure calculated for the period of delay or disruption, and the comparative figures are then used to calculate the amount of loss.'

This, of course, has the merit of simplicity, but I am doubtful about its legal basis. Quantity surveyors asked to approve the use of such a method should, in my view, obtain the specific authority of the employer to settle on this basis. Legally, it seems to me that it is for the contractor to prove his claim; the duty of the architect or quantity surveyor is to 'ascertain' the amount of direct loss and/or expense. 'Ascertain' is defined in the dictionary as 'find out (for certain), get to know' and is different in meaning from 'assess' or 'estimate'.

Some people argue that the actual financial effect of loss of productivity is difficult, if not impossible, to 'ascertain'. They then argue that in the absence of 'precise ascertainment' a reasonable assessment should be made based upon the balance of probabilities. The argument has some attraction, and in practice such calculations (if *backed by evidence*) may be used. But we should remember that claims are equivalent to damages, and must be assessed in accordance with legal rules.

Financing charges

Whenever contractors gather together, the great debate about interest or finance charges as a head of claim continues. The decision of the Court of Appeal in *Rees & Kirby Ltd* v. *Swansea City Council* (1986) appears to have caused confusion rather than resulting in clarity, and it is as well to see what the case actually decided.

Rees & Kirby Ltd v. *Swansea City Council* appears to be authority for five propositions:

- A contractor's application for reimbursement of direct loss and/or expense under clauses 11(6) and 24 of JCT 63 must make it clear, if that is the case, that an element of the claim is for financing charges. The application must, of course, be read in a commonsense way, as was emphasised in the Court of Appeal, and no particular form of notice is required. But the prudent contractor will be especially careful in the drafting of his application so as to make matters plain.
- The contractor's claim must be made within a reasonable time of the

loss or expense having been incurred.

Nonetheless, even if the claim is not within a reasonable time – and that is always a question of fact – the employer's conduct may preclude him from enforcing his strict legal rights. In *Rees & Kirby* itself, the council could not rely on the delay between practical completion in 1974 and the contractor's formal application in June 1978, because of the lengthy negotiations between the parties which had led the contractor to believe that an *ex gratia* payment might be made.

However, too much reliance should not be placed on this principle – which would take litigation to enforce – and contractors should submit their claims sooner rather than later.

- Financing charges cannot be recovered for periods when non-payment arose from an independent cause which was not direct loss or expense.

  This is the most controversial aspect of the Court of Appeal decision and involves a highly technical legal point. In many cases the doing of an act starts off a chain of events which lead to loss or damage, and lawyers call this 'the chain of causation'. If liability is to be established, the original act e.g., the issue of a variation order, must be connected without interruption to the loss or expense suffered or incurred by the contractor.

  On one view, in *Rees & Kirby* the chain of causation could be said to have been broken by intervening events, thus making the finance charges no longer 'direct' expense. But this was not the view taken by the Court of Appeal, despite the protracted period of negotiations over the *ex gratia* payment: almost three years. The court ruled that the delay in payment did not constitute 'direct loss and/or expense' for this period, since it arose from an independent cause. But the intervening period did not break the chain so as to make the expense 'indirect' and disentitling the contractor to claim.
- In calculating interest which compensates the contractor for having a bigger overdraft than he would otherwise have had, it is proper to take account of the fact that banks charge compound interest with periodic 'rests', i.e., on a quarterly basis since this is the usual practice of banks.

  In other words, *compound* interest is payable, and not merely simple interest. In the event, the contractors were held entitled to their financing charges to cover the period between the breakdown of the negotiations and the date when they appended their signature to the final account *reserving their claim to interest.*
- Financing charges run up to the last application by the contractor before the issue of the relevant certificate. Under the structure of

the claims clauses, there is no cut-off point at the date of practical completion. Finance charges continue to constitute direct loss and/or expense until the contractor's last written application, before the architect issues his certificate. Thereafter, they cease to be payable, because the right to finance charges has merged in the right to receive payment under the certificate.

The principles laid down in *Rees & Kirby* carried a stage further the earlier Court of Appeal ruling in *F.G. Minter Ltd* v. *WHTSO* (1980) and, since in *Rees & Kirby* the contract was in an unamended JCT 63 form, placed the contractor's entitlement to financing charges as a head of claim beyond any doubt. The *Minter* case established the basic rule, namely, that finance charges are an integral and constituent part of the 'direct loss and/or expense'. There is thus no justification for a contractor's proven claim being rejected by architect or employer under JCT contracts.

It can be confidently asserted that read together, the two cases establish the contractor's right to payment of finance charges between the date the loss was incurred and the date of his written application to the architect, provided the claim is made within a reasonable time of the loss or expense having been incurred.

Under JCT 63 terms the period of claim can be extended by subsequent applications or a series of notices. This is not necessary under JCT 80 because of the changed wording of clause 26, which refers to past and future losses - but in all other respects the position appears to be the same.

It is not necessary that the contractor should in fact be in an overdraft situation in order to claim finance charges. If he is in the happy position of being in credit then he is entitled to the interest that he might otherwise have earned. So far as overdraft interest is concerned he can readily obtain the supporting evidence from his bankers, although it seems that he is entitled only to the rate of interest normally paid by someone in his position, and not a specially high rate he might have to pay because of his own credit record.

Lost investment potential may be more problematical to prove, but it should not be too dificult.

'Hudson formula' approved?

Advocates of the formula approach to contractor's claims for overheads and profit during contract prolongation were no doubt delighted by the decision of Sir William Stabb, QC, in *J. Finnegan Ltd* v. *Sheffield City*

Council (1989). The judge, so it is said, adopted the 'Hudson formula' or – more accurately – a variant of it, namely the 'Emden' formula. The 'Hudson formula' actually takes the allowance made by the contractor for head-office overheads and profit in his tender, divides it by the original contract period and multiplies the result by the period of contract overrun. It is set out on page 599 of *Hudson's Building Contracts*, 10th edition.

In fact, although Sir William stated that he 'infinitely preferred' the Hudson formula to the plaintiffs' method of calculation, what he actually used was the *Emden* formula, i.e., overhead and profit based on a fair annual average, multiplied by the contract sum and the period of delay in weeks divided by the contract period.

The case arose out of a JCT 63 contract with the Sheffield City Council for improvement works to 34 houses. There were special conditions in the bills about handing over the houses in batches of four, though the contractor had been asked to tender a lump sum on the basis of specified notional work on 34 unknown dwellings, to be completed within a specified time. As the judge remarked, the JCT contract

> 'subsequently entered into was based upon such unusual conditions in the bill of quantities that it should have been realised by the defendants that it was highly unlikely that the progress and completion of the work would conform to the timetable laid down.'

It did not, and Finnegan became contractually entitled to make claims for direct loss and/or expense. The court had to decide the dates between which the prolongation costs were to be calculated and what, if any, allowance should be made for overheads and profit. Finnegan's labour force was two-thirds sub-contracted, and it was argued that any allowance for overheads and profit should be limited to their direct labour force. In the event, Sir William Stabb, QC, held that the prolongation period started on the expiry of the original contract completion date, the bill provisions about the completion date related to programming only and gave rise to no time-related contractual obligation.

The judge had no doubt about the contractor's entitlement to recover his general overheads and profit during the prolongation period:

> 'It is generally accepted that, on principle, a contractor who is delayed in completing a contract due to the fault of the employer, may properly have a claim for head office or off-site overheads

during the period of delay, on the basis that the workforce, but for the delay, might have had the opportunity of being employed on another contract which would have had the effect of funding the overheads during the overrun period.'

The difficulty lies in calculating its amount and proponents of the *Hudson* formula as a panacea of general application should appreciate the assumptions on which it is based. As the reporters' editorial commentary points out 'it was not apparently argued that the formula might be inappropriate in the circumstances' and the judge rejected the plaintiffs' own different formula calculation as being 'too speculative'. Furthermore, the judgment makes clear that the percentage to be taken is the average percentage earned by the contractor on his turnover as shown by his accounts and not the percentage built into the particular contract by the contractor which is the *Hudson* basis: see *Whittall Builders Co. Ltd* v. *Chester-le-Street District Council* (1985).

In the *Whittall* case it is clear that the formula approach was agreed by the parties as the basis of calculation and, whatever else it decides, *J.F. Finnegan Ltd* v. *Sheffield City Council* does not hold that a formula approach can always be used. The *BLR* commentary puts the point well. By the use of the word 'ascertain', JCT contracts make it plain that it is normally 'the actual loss or expense clearly and directly attributable to the event or events in question' which is what matters.

'It is an inquiry based upon fact and not upon estimates or upon the assumption of notional alternative contracts. Accordingly it should not be assumed that the use in this case of the Hudson formula is indicative of any general judicial approval of it particularly since there appears to have been no objection on the part of the employer to an assessment of loss and expense on that basis.'

John Sims and I make the same point in our discussion of the formula approach in *Building Contract Claims* (2nd edition, 1988, pp. 129–36) and we stand by the views there expressed.

What is of significance about the case is the holding that the allowance for overheads and profit was not limited to the contractor's direct labour force since it was the period of delay which was important.

'The work carried out during the overrun must have consisted of both direct and sub-contracted labour and both must have incurred expenditure on overheads. It is this unfunded expenditure which is the subject of this part of the plaintiffs' claim.'

It was a claim for the funding of overheads which the notional other contract would have provided, as was the case in *Ellis-Don Ltd* v. *The Parking Authority of Toronto* (1978). The pity is that the pros and cons of formulae calculations were not fully argued and so the debate between the opposing sides will continue. For the time being, employers must reject the blanket application of the Hudson or any other formula; hard evidence is required in order to back up a contractor's claim.

The true position with regard to formulae was put inimitably by John Parris in his comment on the case in *Construction Law Digest*, Vol. 6, with every word of which I agree:

> 'Nowhere in the judgment is there a reference to the Court of Appeal decision in *Tate & Lyle Food & Distribution Ltd* v. *GLC* (1982) which appears to lay down the proposition that, in quantifying damages for breach of contract it is for a plaintiff to produce evidence of actual loss and he cannot rely upon notional calculations.
>
> The current case will no doubt be relied upon by claims consultants, one of whom I recently heard say that the official referees accepted the Hudson formula. Hitherto, there has been no reported case in which any of them have done so.
>
> Quite apart from that, there are at least three formulae being advocated for the calculation of head office overheads, all of which produce different figures.
>
> All of them suffer from the fundamental defect that they do not refer to the actual situation of the contractor.
>
> It is, of course, quite true that a judge should use "a broad brush" in evaluating a claim for damages where it would be impossible to evaluate every element, as the Court of Appeal has recently reaffirmed in the *Dominion Mosaic & Tile* case, but all the formulae make quite unwarranted assumptions, for example as to whether the contractor had a realistic opportunity to undertake other work if he had not been delayed.'

I share his view that considerable doubt was cast on the formula approach by the case of *Tate & Lyle Food & Distribution Ltd* v. *Greater London Council* (1982). Tate & Lyle sued the GLC and others for negligence and nuisance arising out of the construction of two new piers for the Woolwich Ferry which resulted in the formation of silt deposits which prevented access to Tate & Lyle's barge moorings. The defendant was liable for its failure to dredge away the deposits of silt and, as a result between 1967 and 1974 Tate & Lyle incurred heavy costs in dredging. The parties agreed the amount of damages to be awarded subject to two points. First, whether it should include an amount for managerial and supervisory expenses incurred by Tate & Lyle in attending to the problems caused by the silting and, second, the amount of interest on the damages.

Tate & Lyle claimed 2.5% for managerial time, on the basis of practice in the Admiralty courts where a percentage figure is allowable because any exact computation of the expense due to the disturbance and extra work caused in the ship owner's office by a collision is almost impossible. That, of course, is not the same as a typical contractor's claim for head office overheads.

Mr Justice Forbes accepted that the plaintiff could properly recover damages for the managerial and supervisory expenses directly attributable to the GLC's failure to dredge the silt, but since Tate & Lyle had kept no records of the time expended, its loss could not be quantified either in cash or as a percentage of the damages awarded. It had, therefore, failed to prove its loss. The judge said:

> 'I have no doubt that the expenditure of managerial time in remedying an actionable wrong . . . can properly form the subject matter of a head of special damage. In a case such as this it would be wholly unrealistic to assume that no such managerial time was in fact expended. I would also accept that it would be extremely difficult to quantify. But modern office arrangements permit of the recording of time spent by managerial staff on particular projects . . . While I am satisfied that this head of damage can properly be claimed, I am not prepared to advance into an area of pure speculation when it comes to quantum.'

This, in fact, sounded the death knell for any *theoretical* formula approach, but if the formula can be substantiated from the contractor's records, the position is different. What *Tate & Lyle* appears to establish is that a claim for overheads must be precise and capable of proof and withstand critical examination. It must not be an assumed or theoretical amount.

If the contractor can produce *acceptable* evidence for his head office overheads claim he is entitled to claim for them. The problem is, of course, that the head office overheads allowance built into the tender may never in fact have been achievable – perhaps due to the contractor's own inefficiency.

Two possible ways of establishing such a claim are:

- The cost of the head office resources. This must be done by calculating the actual amount incurred due to the events giving rise to the claim, and this assumes proper and adequate records.

- Alternatively, he must establish that resources were kept on site (where there is a prolongation situation) and were thus prevented from earning a contribution by working elsewhere.

There are practical difficulties in both approaches, but the task is not impossible, and in *Ellis-Don* v. *The Parking Authority of Toronto* (1978) an approach based on the Hudson formula was accepted. But it was only accepted on the basis that the contractor was able to establish with hard evidence that had he not been delayed on the particular contract he would have been able to recover his overhead contribution elsewhere.

The court also required evidence that the delay prevented the contractor from using his resources elsewhere and the well-known English case of *Peak Construction (Liverpool) Ltd* v. *McKinney Foundations Ltd* (1970) establishes that the latter holding is also the law of England.

Contractors, architects, engineers and quantity surveyors alike should remember, of course, that what is required is proof 'on the balance of probabilities' and not the criminal standard of proof.

But architects and engineers - like judges and arbitrators - cannot embark on pure speculation when it comes to the quantification of any claim, whether for head office overheads or any other permissible head of claim.

Interest as 'special damages'

The controversial topic of interest or financing charges on late payments was thrown into the melting pot by the decision of Judge Lewis Hawser QC, then Senior Official Referee, in *Holbeach Plant Hire Ltd* v. *Anglian Water Authority* (1988).

The case was an appeal from an arbitrator's award given in a dispute arising under the ICE Conditions of Contract, 4th edition, in an amended form. The contractor contended that the engineer ought to have included sums in his certificates for 'financing charges' for late certification. He claimed this as 'special damages', i.e. loss of a kind which the law will not presume in a claimant's favour, but which must be specifically pleaded and proved.

In fact, clause 60(6) of the ICE Conditions, 5th edition, gives a contractual right to interest on overdue payments. However, that clause does not appear to have been relied on in the case as reported. Judge Hawser defined the issue before him in this way:

'Assuming (1) the applicants can establish that they have sustained loss by way of interest or financing charges as claimed; (2) such loss was caused by the respondent's default; and (3) the respondent had knowledge of facts or circumstances which made such loss a not unlikely consequence of such default; whether the applicants are entitled to recover such loss . . . as special damages.'

He answered that question affirmatively and the case was remitted to the arbitrator. He had concluded, perhaps not surprisingly, on the cases cited to him that financing charges were a normal part of most construction contracts and so must be deemed to be within the parties' contemplation as a natural result in the ordinary course of things from the breach of late payment. On that basis, they only entitle the claimant to general damages. These cannot include interest.

This has been the law since the case of *London Chatham & Dover Railway Co.* v. *South Eastern Railway Co.* (1893). Statute apart, interest cannot be recovered as damages for late payment unless:

- The contract specifically provides for the payment of interest, thus creating a contractual right;
- It forms a constituent part of the debt, i.e. is part of 'direct loss and/or expense': *F.G. Minter Ltd* v. *WHTSO* (1980); *Rees & Kirby Ltd* v. *Swansea Corporation* (1983);
- The claim is specifically pleaded and proved as special damages, as in *Wadsworth* v. *Lydall* (1981). In that case, the plaintiff contracted to sell a property to the defendant and in anticipation of receiving the purchase price, contracted to buy another property. The defendant buyer defaulted. As a result the plaintiff had to pay his vendor interest on the unpaid purchase price of the other property and incurred mortgage costs to raise the balance. The Court of Appeal allowed the interest which he had to pay his vendor for late completion and the costs of raising the mortgage.

Holbeach Plant Hire Ltd v. *Anglian Water Authority* supports the view expressed by the writer and John Sims in *Building Contract Claims*, 2nd edition, p. 154, that *Wadsworth* 'seems . . . to open up the way to such claims for interest as "special damage" in arbitration proceedings under building contracts generally.' The position under JCT contracts is clearly dealt with by *Minter*.

The common law position was, of course, restated in a modern and restricted form by the House of Lords in *La Pintada* (1984). The House of Lords considered the matter further in *President of India* v. *Lips Maritime Corporation* (1987). In *Holbeach*, Judge Hawser adopted some propositions put forward by Lord Justice Neil in the Court of Appeal judgment in *Lips* which, while overuled by the House of Lords, was not reversed on the matter of interest on overdue accounts.

Lord Justice Neil put forward three propositions:

- A payee cannot recover damages by way of interest merely because money has been paid late. The courts will not impute to the parties knowledge that in the ordinary course of things late payment will result in loss. The reason for this appears to be that there is a statutory right to interest in specified circumstances.
- A payee can recover damages for late payment if he can prove facts which make the damage 'special damage' under the second part of the old rule in *Hadley* v. *Baxendale* (1854). This is where the loss is 'such as may reasonably be supposed to have been in the contemplation of both parties at the time they made the contract, as the probable result of the breach of it'. This is at least arguably the case in almost every construction project.
- In every case, what must be determined is the loss which was reasonably within the contemplation of the parties at the time the contract was made. A plaintiff will be able to recover damages in respect of a special loss if it is proved that the parties had knowledge of facts or circumstances from which it was reasonable to infer that delay in payment would add to that loss', said Lord Justice Neil.

Holbeach Plant Hire Ltd v. *Anglian Water Authority* is an exciting case and quite consistent with the authorities. Even in its restricted form the *London Chatham & Dover* rule is commercially unjustifiable. In *La Pintada* the House of Lords would not change the law because Parliament had done so to a more limited extent than most of us would have liked. If payment is due on 1 August, payment made on 1 October is not the same thing; the payee is out of pocket. Far too many employers do ignore the contractual terms for payment.

In most cases the matter is forgotten. Now if it comes to litigation or arbitration, the loss can be claimed as 'special damages' which – subject to proof – may give a better and more realistic return than the statutory right to interest. The most satisfactory solution would be further legislation, but there is little hope of that.

Who pays the claims consultant?

Under neither JCT nor ICE contracts is the cost incurred by the contracter in preparing a 'claims submission' recoverable as part of 'direct loss and/or expense' or 'cost' or 'expense'. *James Longley & Co. Ltd*

v. *South-West Thames Regional Health Authority* (1983) is the case quoted to support the contrary view, but it was about the fees paid to a claims consultant in an arbitration. My view that the ruling is of limited application is regularly challenged at seminars, the suggestion being that if the contractor employs a specialist consultant to prepare his claim against the employer, the fees involved become part of the contractor's 'direct loss and/or expense', etc.

That this is not so is supported by the judgment itself, which contractors and claims consultants ought to read. All that the case establishes is that consultants' fees for work done in preparing the case for arbitration are allowable (or may be allowable), *as those of an expert witness*. The consultants' fees in respect of work done in preparing the contractor's final account for submission to the architect and for work done in the course of the arbitration as general adviser were expressly disallowed by the taxing Master, and were not at issue. They are not allowable.

The claim arose out of a JCT 63 contract for the construction of an extension to Worthing Hospital. The contract work should have been completed by 22 November, 1973. The works were not, in fact, completed until 8 February, 1975. Longley claimed extensions of time under clause 23, as well as 'direct loss and/or expense' under clause 11(6). It submitted its claim to the architect, but was dissatisfied with his decision, and the matter went to arbitration.

The arbitration hearing was expected to last 16 weeks, but after 16 days the parties came to terms. The final award was by consent, and the arbitrator ordered that the employer should pay the contractors £136,000 plus costs. These included an amount of £16,022 for fees paid to a claims consultant, only part of which was allowed (£6452), when the bill was taxed in the High Court. (Taxation has nothing to do with income tax; it is a process whereby a High Court official, called the Master, assesses whether the costs are reasonable and decides which part should be paid by the unsuccessful party.)

The amounts disallowed related to the consultants' fees in respect of work done in preparing the claim itself and work done as general adviser. The Master agreed, despite the respondents' objections, that the £6452 in respect of fees for work done in preparing the case for arbitration, after the rejection of the initial claim, were allowable as costs of an expert witness.

The RHA appealed against this ruling, but its appeal was dismissed by Mr Justice Lloyd – a very experienced commercial judge. RHA argued that the consultant was not professionally qualified and was, therefore, not an expert. 'That argument is groundless,' said the judge. 'An expert may be qualified by skill and experience, as well as by professional qualifications' and he found and held that there was no reasonable doubt as to the expert's skill and experience.

The second line of attack was that the consultant would never have given evidence at the hearing – no expert report had been exchanged prior to the hearing as is usual in the case of expert witnesses. That argument disappeared in face of Counsel's statement that the consultant would have been called to prove the Schedules which he had prepared.

RHA's main point was that the consultant's evidence would have been inadmissible because it would have gone to the very point which was before the arbitrator. An expert may not be asked the question which the arbitrator himself has to decide. 'The dividing line between what an expert witness can and cannot be asked is often very narrow,' said the judge.

The case involved the interrelation of various overlapping causes of delay and disruption, and Mr Justice Lloyd was of the opinion that the arbitrator – or a judge – would have been helped by expert evidence. 'The assessment of delay resulting from any individual cause called for a high degree of skill, as well as experience of how in fact building operations are carried on.' Far from lengthening the hearing, the expert evidence might well have the opposite effect. In a complex building dispute – even before an expert arbitrator – such evidence might well simplify the issues and save time and money.

An arbitrator, even though he is himself an expert, can still consult other experts. 'One reason for having an expert arbitrator is so that he should be able to understand the expert evidence, not so that he should do without it,' said the judge and, in his view, the consultant's fee was properly allowable.

This is far from saying – as some people have alleged – that if a contractor employs a claims consultant to prepare his case, that person's fees are allowable as part of 'direct loss and/or expense' under JCT 80 or 'expense' under ICE Conditions. The *Longley* case goes against this view.

Under JCT and other standard forms of contract the contractor is not entitled to reimbursement for consultant's costs which he has incurred in preparing the claim. JCT terms do not, in fact, require the contractor to prepare a claim, but merely to submit substantiating evidence for the architect or quantity surveyor to consider in making the ascertainment; the position is similar under an ICE Contract.

Fees paid to outside experts, such as quantity surveyors, to prepare a claim are not allowable. The position is different where the claim is refused and proceeds to arbitration or litigation. This does not shut-out 'claims costs' entirely because the expenditure of managerial time (over and above the norm), as a result of the event relied on, is clearly the subject-matter of a claim. That is assuming that this time and cost can be evaluated and supported by evidence and is not covered already by a claim for head office overheads: *Tate & Lyle Ltd* v. *GLC* (1982).

But the general rule remains – and is emphasised by the *Longley* case – that the contractor is not entitled to the costs of preparing the claim, since this is not something the contract requires him to do.

The position on interest

The House of Lords stilled the long-standing debate about interest through its ruling in *President of India* v. *La Pintada Cia Navegacion SA* (1984) and restated the old common law rule. In the absence of agreement between the parties regarding payment of interest on a debt due, late payment of debt does not attract interest as general damages.

This rule applies to late payment under JCT 80, for example, because the contract makes no provision for payment of interest on sums paid late. In contrast, clause 60(6) of the ICE Conditions of Contract, 5th edition, gives a specific right to interest for either delay by the engineer in certifying, or late payment by the employer, though at a very modest rate.

Where a dispute is taken to court or arbitration, the position is different. Section 35A of the Supreme Court Act 1981 (and corresponding provisions in the County Courts Act 1959 and the Arbitration Act 1950), confers a discretionary power to award interest in two cases:

- Where a debt is paid late, after proceedings for its recovery have begun, but before they have been concluded.
- Where a debt remains unpaid until judgment or award.

It should be noted, however, that neither the courts nor arbitrators can award *compound* interest; only simple interest is allowable. The House of Lords has overruled the case of *Tehno-Impex* v. *Gebr Van Weelde* (1981), which is now of no relevance, either in the case of contractor's and/or expense claims or otherwise.

The situation, was, however, complicated by the decision of the Court of Appeal in *Wadsworth* v. *Lydall* (1981), which has been approved by the House of Lords. The rule is that if a creditor can prove that he has suffered special damage, e.g. by himself having to pay overdraft interest, as a result of the debtor's late payment, the creditor is entitled to claim the interest paid as special damages – even if the debt is paid before proceedings for its recovery are started – provided the debtor knew, or ought to have known, that late payment would involve the creditor in expense by way of financing charges.

The distinction between general and special damages is the difference between damages recoverable under the first part of the rule in *Hadley* v. *Baxendale* (1854), which are general damages, and those recoverable under the second part of the rule (special damages) by reason of special matters known to both parties at the time of contracting. The rule in *Wadsworth* v. *Lydall* in fact considerably attenuates the harshness of the common law rule against interest.

In the construction industry, it is now clear and settled law that interest or financing charges are recoverable under the 'direct loss and/or expense' provisions of the JCT forms, and this extends to loss of interest which might have been earned by investment. The authority for this proposition is the Court of Appeal decision in *F.G. Minter Ltd* v. *Welsh Health Technical Services Organisation* (1980), even disregarding the gloss put on that case by *Rees & Kirby Ltd* v. *Swansea Corporation* (1983).

In the *Minter* case, the contract was in JCT 63 terms and clauses 11(6) and 24(1) had been amended so as to require the contractors to make written application within 21 days of it becoming apparent that progress was materially affected, instead of requiring this to be done 'within a reasonable time'. I have heard it argued that this in some way limits the *Minter* principle, but this view is not well-founded. What is being claimed in these circumstances is not interest on a debt, but a debt which has, as one of its constituent parts, interest charges which have been incurred: see the views of Lord Justice Ackner in the *Minter* case.

In *Building Contract Claims*, 2nd edition, (BSP Professional Books) 1989, the writer and John Sims discuss some of the practical problems involved in the application of this ruling: see pp. 152–7. Those pages may need some revision in light of later developments. In a 'direct loss and/or expense' the architect or quantity surveyor must include an element of finance charges in ascertaining the amount. The starting point must obviously be the interest actually payable to the claimant contractor, or the interest he would otherwise have earned by ordinary investment, disregarding, any special attributes of his.

However, any special position which the contractor is in must be disregarded, e.g., if he is paying a higher rate of interest than the average contractor: see the views expressed by Mr Justice Forbes in *Tate & Lyle Ltd* v. *Greater London Council* (1982). Proof would be by way of an auditor's or banker's certificate. Some contractors may waver from a borrowing to a lending position during the period of the contract, and this can cause complications in the ascertainment. In such cases it is thought that a general average over the period can or should be taken.

Under JCT 80 (and under the Intermediate Form), in practice, the element of finance charges forms a very small part of any claim, whereas under JCT 63 terms it could be a substantial element.

The period of time involved begins to run from the day the loss is incurred, which may or may not coincide with one of the events giving rise to the claim. The architect or quantity surveyor is '*to ascertain the amount of such loss and/or expense* which has been, or is being incurred,' and amounts so certified in whole or in part, are to be taken into account in the next interim certificate: clause 26.5. If this provision is properly and fairly operated, the employer's possible liability for financing charges as an element of claim should be reduced. But when so included they must be calculated up to the date of the ascertainment.

Chapter 8

Set-off and Counterclaim

Rights of set-off

Where a building contract or sub-contract contains a set-off clause, it entitles employer or contractor to off-set claims for disruption and delay against sums certified as due to contractor or sub-contractor. The effect of these clauses depends entirely on how they are worded.

Any equitable or other right of set-off can be excluded by an appropriately-drafted contract term and probably by necessary implication by looking at the contract as a whole. Indeed, in construction contracts the express contractual right of set-off normally excludes any rights that would otherwise be implied, and it is, therefore, essential to follow the contract procedures exactly.

These principles flow from the decision of the House of Lords in *Gilbert Ash (Northern)* v. *Modern Engineering (Bristol) Ltd* (1974), which clearly overruled the line of cases beginning with *Dawnays Ltd* v. *F. G. Minter Ltd* (1971) where the Court of Appeal had seemingly held that certified sums were the equivalent of cash or, as was said, bills of exchange. The *Gilbert Ash* case was followed in the House of Lords by *Mottram Consultants Ltd* v. *Bernard Sunley & Sons Ltd* (1975). Following the *Gilbert Ash* case, the industry was in a turmoil because certificates became of limited value where the assertion of a right of set-off postponed the payee's right to immediate postponement.

The negotiating bodies, therefore, dealt with the situation by amending the various contract and sub-contract forms so as to provide an express right of set-off, subject to certain conditions. This should, in fact, make for simplicity. That it does not do so, is clearly shown by a recent set-off case which lays down new ground rules and emphasises the importance of the wording of the set-off clause.

The case is *Chatbrown Ltd* v. *Alfred McAlpine Construction Southern Ltd* (1986), a decision of His Honour Judge James Fox-Andrews, QC, an official referee, and it involved the well-known NFBTE/FASS Standard Form of Non-Nominated Sub-Contract in its July 1978 revision. Chatbrown issued a writ for some £224,000 representing the value of measured work and sought summary judgment. McAlpine agreed that for the purposes of the case they owed Chatbrown £231,480, less the amount of their set-off, which was an equal amount, and applied for the proceedings to be stayed and go to arbitration.

The case turned on the interpretation of clause 15(2) of the sub-contract. This provides that:

> 'The contractor shall be entitled to set-off against any money . . . otherwise due . . . the amount of any claim for loss and/or expense which has been actually incurred by the contractor by reason of any breach of, or failure to observe the provisions of this sub-contract by the sub-contractor . . .'

This right is subject to several provisos.

McAlpine had served the necessary notice under the clause on the sub-contractors, and it included a claim for an item called 'Assessment of additional costs'. These were in respect of moneys which would be laid out at a future date resulting from alleged past delays of Chatbrown.

McAlpine's point was that these assessed costs constituted a present loss or expense at the time the notice was given. It was argued that 'incurred' in clause 15(2) meant 'incurred a liability for loss and/or expense' and that the word 'actually' added nothing to the meaning of the phrase.

Readers may be surprised that the point was litigated – as I was myself – but now we have an authoritative ruling on what has been a very contentious point. The judge pointed out that there were four main requirements of the notice under clause 15:

- The claim to set-off must relate to loss and/or expense actually incurred.
- It must have been quantified in detail.
- It must have been quantified with reasonable accuracy.
- Notice must have been given at least 17 days before amounts would otherwise become due under the sub-contract.

He only had to consider the first point, and in so doing he reviewed a number of authorities and considered the dictionary meaning of 'actually' and 'incur'. The starting point was to consider what the parties had meant by using the qualification 'actually' and the word 'incurred'.

Past case law suggests that 'incurred expenses' and similar phrases, at least meant that those expenses had been paid, or at least the payor had become liable to pay them as distinguished from estimated expenses. 'Incurred' suggests expenses, etc., which have been paid out.

After reviewing the authorities, Judge Fox-Andrews concluded:

> 'It is always necessary to be cautious before interpreting words appearing in one document or statute on the basis of decisions on similar words appearing in other contexts. However, I am satisfied that meaning (i.e., that stated above) should be attached to the word "actually". It appears to me that it is intended to add emphasis to the fact that loss and/or expense had been incurred and, further, it has a temporal connotation.'

Accordingly, he ruled against McAlpine and gave summary judgment for Chatbrown. He was certain that the 'loss and/or expense' did not include McAlpine's assessed figures

This judgment is consistent with authority and, while it is in theory applicable only to the blue form of sub-contract, it can be taken as representing the position where like words are used in other set-off clauses. The right of set-off conferred on the main contractor by clause 15 of the blue form of domestic sub-contract does not extend to loss and/or expense not incurred at the date of the contractor's notice. It certainly does not cover assessed future costs - even if quantified with reasonable accuracy. Loss and/or expense must have been incurred, i.e. moneys must have been paid out. The fact that future liabilities may accrue is quite immaterial. I think that the decision is sound law.

Common law rights: hard to shake off

The main contractor's right to set off any claim for damages at common law is not excluded by clause 15(3)(d) of the FCEC Form of Sub-Contract for Civil Engineering Works. Very clear words are necessary to exclude the common law right to set-off.

This was the ruling of the Court of Appeal in *NEI Thompson Ltd* v. *Wimpey Construction UK Ltd* (1987). The dispute arose out of the construction of 2.7 km of roadway in West Glamorgan, where Wimpey was the main contractor and NEI Thompson sub-contractor for the supply and erection of structural steelwork for a number of bridges. The sub-contract was in FCEC form and for £444,519.

The sub-contract was for 11 weeks but was completed nine weeks late. In August 1986 NEI Thompson submitted an application for an interim payment amounting to £157,186. Wimpey declined to make any payment on the grounds that it had a cross-claim for damages for delay, later calculated as being at least £458,957. NEI Thompson's case was that Wimpey was too late to rely on its cross-claim because it had not given notice under clause 15 in time. If that was correct, Wimpey would have to pursue the damages claim by arbitration as provided by the sub-contract.

The case turned mainly on the wording of clause 15(3)(d):

> 'In the event of the contractor withholding any payment he shall notify the sub-contractor of his reasons in writing as soon as is reasonably practicable but not later than the date when such payment would otherwise have been payable.'

The case raises the familiar question of set-off, on which the leading cases are now *Gilbert Ash (Northern) Ltd* v. *Modern Engineering (Bristol) Ltd* (1974) and *Mottram Consultants Ltd* v. *Bernard Sunley & Sons Ltd* (1975). These House of Lords' decisions sounded the death knell of the so-called principle of *Dawnays Ltd* v. *F.G. Minter Ltd* (1971).

NEI Thompson strongly relied on something said by Lord Diplock in the *Gilbert-Ash* case. In interpreting a construction contract

> 'one starts with the presumption that neither party intends to abandon any remedies for its breach arising by operation of law, and clear express words must be used in order to refute this presumption'.

NEI Thompson argued that Lord Diplock's statement must be read in light of what Lord Cross said in *Mottram* v. *Sunley*:

> 'The position is now once more what it was before *Dawnays'* case was decided – namely that one should approach each case without any preconception about the existence of a right of set-off, though one must bear in mind the principle established in *Mondel* v. *Steel* (1841).'

NEI Thompson criticised the trial judge on the basis that he started with a presumption in favour of allowing a set-off. It said there is no presumption either way. The Court of Appeal decisively rejected this argument.

Lord Justice Lloyd quoted what the trial judge had said:

> 'It seems to me that if a right which a party would otherwise have is to be excluded by a contract, one has to see whether that contract has clear words of exclusion . . . Although clause 15(3)(d), taken by itself, is wide enough to cover the circumstances in which payment

was withheld, it is not clear enough in the context of clause 15(3)(b) [which entitles the contractor to withhold or defer payment in specified circumstances "without prejudice to any rights which exist at common law"] to exclude the contractor's right of set-off'.

The Court of Appeal thought that this approach was correct and that at first sight the judge was right. NEI Thompson said that 'any payment' in clause 15(3)(d) must mean 'any payment', including any payment withheld by Wimpey in the exercise of its common law right to set off its claim for damages. There was nothing to limit the scope of the clause.

Wimpey said that looking at clause 15(3) as a whole it was clear that the obligation to notify reasons under clause 15(3)(d) within a specified time referred back to the withholding of payment under specific provisions of clause 15(3). All of these relate to amounts or quantities included in the sub-contractor's application for interim payment. The Court of Appeal agreed with Wimpey's contention.

To quote Lord Justice Lloyd:

> 'If any question arises in relation to those amounts or quantities, then the [contractor] must give notice within 35 days, as provided by clause 15(3)(d) in order to justify withholding payment. But it would require much clearer words . . . to exclude the contractor's ordinary right of common law to rely on set-off in respect of some dispute which has no connection with amounts or quantities.'

The inclusion of the words 'without prejudice to any rights which exist at common law' in clause 15(3)(b) reinforced this conclusion. They made it quite clear that the right of set-off at common law was intended to be completely outside the machinery of clause 15.3. Not only was clause 15(3)(d) insufficiently clear to exclude the common law right of set-off, but that right was specifically preserved by the earlier sub-clause.

The position is entirely different under the JCT Nominated Sub-Contract, NSC/4. Clause 23.4 of that document provides that

> 'the rights of the parties . . . in respect of set-off are fully set out in the sub-contract . . . and no other rights whatsoever shall be implied as terms of the . . . sub-contract relating to set-off'.

That is certainly clear enough to exclude common law rights, even if the wording is somewhat clumsy and inelegant, but as the Court of Appeal decision in *Acsim (Southern) Ltd* v. *Danish Contracting & Development*

Co. Ltd (1990) shows, the set-off provisions do not contain the only machinery whereby a sub-contractor's right to payment can be challenged: see p. 134.

Set-off under NSC/4

The complexity of the set-off provisions of the various standard sub-contract forms has led to a good deal of litigation and the cases emphasise that the formalities must be complied with strictly if a set-off is to be valid. The majority of cases has involved the older standard forms of sub-contract sponsored by NFBTE/FASS.

William Cox Ltd v. *Fairclough Building Ltd* (1988) is about the set-off provisions of NSC/4 – the JCT Standard Form of Nominated Sub-Contract – and emphasises the importance of the issue of the architect's certificate under clause 35.15.1 of JCT as a pre-condition to the main contractor's right to claim against the sub-contractor. 'Where the contractor has to rely on a certificate', said Judge James Fox-Andrews, '[it] must have been issued prior to the date on which the moneys become due.' If it is not, then the purported exercise of the right of set-off will be invalid.

In *William Cox Ltd* v. *Fairclough Building Ltd*, Cox was nominated sub-contractor for glazing works at a sports centre at Chesterfield. They claimed summary judgment under Order 14 of the Rules of Court for amounts of £8345.06 and £5791.14, plus interest, in respect of sums certified as due to them. Fairclough's broad contention was that they had a set-off or counterclaim exceeding Cox's claim and that consequently Cox was not entitled to judgment in any sum. As the sub-contract contained an arbitration clause, they asked for the proceedings to be stayed under section 4 of the Arbitration Act 1950.

NSC/4, clause 23, confers an express right of set-off on the main contractor in respect of the 'amount of any claim for loss and expense which has actually been incurred' as a result of delay, etc., by the sub-contractor, subject to various provisos as to quantifying its amount in detail and with reasonable accuracy and the giving of appropriate notice of intention to set off at the right time.

Fairclough alleged that Cox was in delay and, on 28 January 1987, gave Cox notice of alleged failure to complete in time and notified their intention to set-off an amount of £35,551, which was quantified in detail. They sent a further letter and notice on 12 March 1987, 'which supersedes our earlier assessment of 28 January 1987', the total amount by this time being £66,945, including a sum for scaffolding expenses.

Fairclough confirmed the scaffolding costs in a letter to Cox dated 17 April 1987, a month after the architect had issued a certificate showing moneys due to Cox, and further certificates were issued in May and June, the effect of which was that a single sum of £5791.14 was payable to Cox on 27 June 1987.

The architect's certificate of delay was dated 4 June and was stated to be 'without prejudice to [Cox's] claim for extension of time'. The sub-contract works were practically completed in their entirety on 4 September 1987.

Despite certain conflicts in the documentation, the learned judge was satisfied that Fairclough had made out a *prima facie* case that they had actually incurred £4564 scaffolding expense by 12 March and, since payment of the first certificate only became due on 3 April (17 days from issue of the certificate) the letter of 12 March was a valid notice of intention to set off; he reached the same conclusion as to the letter of 7 April.

Cox contended, however, that the architect's letter was not a valid certificate of delay because of its reference to pending applications for extensions of time. Judge Fox-Andrews sagely remarked that 'the provisions of the main contract and the sub-contract relating to his kind of certification are not too easy to interpret'. But he looked at the realities.

> 'If a valid certificate cannot be granted until the last [application by a sub-contractor for extension of time] it would be simple for a sub-contractor by constant application to prevent an effective certificate being issued until a very late stage.'

He ruled that the letter was a valid certificate and, since it entitled Fairclough to set off the expenses detailed on 7 April, Cox's claim for summary judgment failed in that respect.

There remained the question of moneys due under the earlier certificate, since no certificate of delay had been issued on 3 April when the moneys became payable to Cox. The judge acceded to the argument that the certificate must have been given by the date the moneys become due under the terms of the contract. It is not sufficient that the certificate of delay be issued later, as was the case here. In the result, he adjudged that Cox was entitled to judgment for £8345.06 and interest amounting to £1536.64, the balance of the claim being stayed to be arbitrated.

This important judgment provides welcome clarification of the NSC/4 set-off provisions and emphasises that the interpretation of a commercial contract must yield to commonsense. It is the first direct authority on the point since, as the judge pointed out, the well-known case of *Tubeworkers Ltd* v. *Tilbury Construction Ltd* (1985) was not relevant. In that case it appears that there had been no quantification of the set-off or notification not less than 17 days before the moneys became due as required by the not dissimilar provisions of clause 13A(2)(c) of the old 'Green Form'.

There is no doubt the complex certification and set-off procedures will produce more case law before long.

Set-off: a total remedy?

The controversial ruling in *BWP (Architectural) Ltd* v. *Beaver Building Systems Ltd* (1988) that the set-off provisions in the industry's standard forms of sub-contract contain the exclusive machinery by which the main contractor can challenge the sub-contractor's right to an interim payment was squashed by the Court of Appeal in *Acsim (Southern) Ltd* v. *Dancon, Danish Contracting & Development Co. Ltd* (1990), which restores a semblance of order to the sub-contract payment regime.

Acsim were domestic sub-contractors for mechanical installation work at the new Scandic Crown Hotel in London. Dancon were both building owner and main contractor. The sub-contract incorporated the terms of the old NFBTE/FASS Blue Form of Sub-Contract (Revised July 1978), clause 13 of which entitled Acsim to monthly interim payments comprising 'the total value of the sub-contract works properly executed' (subject to retention and any discount) within 14 days of application. Clause 15 was the set-off clause which required proper notice of set-off to be given by the contractor before payment was due.

Acsim applied for and received various interim payments. On 27 August 1988 Acsim sent to Dancon an interim application in the sum of £221,018.03, calculated on the basis that they had completed 98% of the sub-contract work. It was common ground on the appeal that any sum due from Dancon to Acsim under that application became payable on 15 September 1988 and that Dancon had not given a set-off notice in time. Dancon did not pay any part of the sum claimed, but raised various claims by letter of 3 October. Acsim issued a writ and applied for summary judgment for £221,018.03 asserting that Dancon could have no defence to the claim because of the provisions of clause 15.

Acsim's application came before his Honour Judge Lewis Hawser QC, the Senior Official Referee, on 24 November 1988. The only evidence of any defence available to Dancon before him was Dancon's letter of 3 October and an affidavit of Dancon's project manager which asserted numerous breaches of sub-contract by Acsim, all of which related to cross-claims caught by the provisions of clause 15, but which disputed Acsim's contention that Dancon had no defence because he alleged that the works had not been properly executed within the meaning of clause 13(2) and that Acsim had not provided an adequate breakdown of the amounts claimed.

The judge refused an application by Dancon for a short adjournment

> 'because the defendants did not give notice of set-off in accordance with clause 15, [he] did not consider that the defendants had any defence to the plaintiff's claim and saw no reason to adjourn the hearing to enable further evidence, including expert evidence, to be submitted.'

He gave summary judgment £221,018.03 with interest.

Dancon appealed saying that the judge had been wrong to refuse an adjournment because the further evidence was irrelevant. They said it would have shown that Acsim were only entitled to £36,952.

The Court of Appeal allowed the appeal. They held that the set-off provisions in clause 15 did not contain the only machinery whereby the contractor can challenge a sub-contractor's right to an interim payment. Clause 15 contains only the rights of the parties in respect of set-off. It does not affect the contractor's right to defend a claim for an interim payment by showing that the sum claimed includes sums to which the sub-contractor is not entitled under the terms of the sub-contract or to defend it by showing that, by reason of the sub-contractor's breaches of contract, the value of the work is less than the sum claimed. The contrary statements in the *Beaver* case (1988) were wrong.

On the proper interpretation of the sub-contract, the defendants had not lost the right to dispute the amount of an interim application by showing that it included the value of work not in fact done or was calculated on an incorrect basis or that part of the work done was worth less than the agreed price by reason of breach by the sub-contractor of the terms of the sub-contract. Accordingly, the learned judge erred in law in holding that the defendants' failure to give notice of set-off under clause 15 deprived them of any defence in respect of the matters raised by them to the effect that the amounts claimed were not properly due.

On balance, the right course was for the Court to give effect to the further evidence and allow the appeal. Judgment for £36,952 (the amount admitted due) was given, and Acsim were directed to repay the amount overpaid.

I do not suppose that this case will still the great ado about set-off clauses even though it applies to the various other similarly worded sub-contracts, including NAM/SC with which the *Beaver* case was concerned. However, it does seem to be a commonsense decision.

The Court of Appeal pointed out that the clear and express words of any contract can in fact provide that the party promising to pay instalments as the work is done abandons his right either to show that the other party has not in fact earned all the sum claimed, or his right to defend the action on other grounds. The sub-contract at issue had not done so, no more than any of the other forms do. All grounds for contesting or reducing the sum claimed can, by suitable words, be made subject to a special regime, but clause 15 and its clones are not sufficiently plain to have that effect. It is good that sanity has returned to this controversial area.

Pay when paid?

There are a number of interesting cases from the Far East which are relevant when considering 'pay when paid' clauses under English law. The Hong Kong and Singaporean sub-contracts are largely based on English models, and the applicable legal principles are the same. For this reason the Far Eastern rulings are of more than persuasive authority in the United Kingdom, especially when the wording of the contract clause at issue is on all fours with its English counterpart.

In *Brightside Mechanical & Electrical Services Group Ltd* v. *Hyundai Engineering & Construction Co. Ltd* (1988) clause 11(b) of a sub-contract provided that:

> 'Within five days of the receipt by the contractor of the sum included in any certificate of the architect the contractor shall notify and pay to the sub-contractor the total value certified therein . . . less . . . any sum to which the contractor may be entitled in respect of delay and completion of the sub-contract works or any section thereof.'

The plaintiffs were mechanical services sub-contractor to the defendant main contractor. The architect certified a sum of some S$1.6 million as being due to the plaintiffs which, after the allowable deductions, came to S$924,711. Owing to delays in completion of the project, the employer declined to make any payment to the defendant main contractors, holding the money due to be set-off against liquidated damages to be paid out for the delay, the architect having issued the relevant main contract certificate of delay. In turn, the main contractors refused to pay the S$924,711 sum over to the plaintiffs, who then sued the defendants and claimed summary judgment.

The main contractors contended that as they had not received any money from the employer they were not liable to pay anything to the sub-contractor. The Registrar found in favour of the defendants, and ordered the proceedings be stayed pending arbitration. He made no order on the sub-contractor's application for summary judgment. The sub-contractor appealed to the High Court.

Mr Justice Thean dismissed the sub-contractor's appeal. In his provisional view, *prima facie* clause 11(b) contemplated the actual receipt by the main contractor of the sum included in the certificate and so until the main contractor actually received from the employer the sum claimed by the sub-contractor, the main contractor was not obliged to pay the money over. In this case, he said, there was no actual receipt by the defendants of the amount certified by the architect.

Sub-contractors working under similar pay when paid clauses should not be too disheartened by the decision as the editorial commentary to the report makes clear.

> 'It is always a question of the proper interpretation of the contract whether [such a clause] is intended to apply regardless of the reasons why payment has not been received by the main contractor . . .'

In this case, curiously, the main contractor had not obtained an architect's certificate against the sub-contractor under clause 8(a) of the sub-contract (equivalent to the old clause 27(d)(ii) certificate) so as to give the main contractor monetary rights against the sub-contractor in respect of alleged delays. Such a certificate is a precondition to the main contractor's right to recover damages against the sub-contractor: *Brightside Kilpatrick Engineering Services Ltd* v. *Mitchell Construction Ltd* (1975). To avoid this requirement, the main contractors argued that they could set-off any sum they might be liable to pay to the sub-contractors on the basis that the plaintiffs might be liable to indemnify them for any delay by them, and this point of law, although arguable, was in fact referred to arbitration.

The real difficulty in the decision, as the *Building Law Report* editors note, is that it is difficult to read a clause like this one 'as applicable where the reason for the failure of the contractor to receive the money is a matter for which he alone is apparently responsible, namely, in this case, his failure to complete the main contract works on time'.

The matter has been canvassed in several Hong Kong cases cited by Mr Justice Thean, but has not been directly decided. Quite clearly, I think, as a matter of general law the employer's obligation to pay a main contractor a certified amount can be discharged by the employer establishing a right to liquidated damages which extinguishes the main contractor's right to payment, but it seems to me to be hard law if a sub-contractor is deprived of his right to payment for work done properly and on time merely because of some breach by the main contractor of his main contract obligations.

Clause 11(b) of this sub-contract referred to 'the receipt by the contractor of the sum included in any certificate of the architect' and these words, in Mr Justice Thean's view, were 'clear and unambiguous . . . giving them the ordinary or normal meaning, they contemplate actual receipt by the main contractor of the sum included in the certificate'. Literal interpretation must sometimes give way to commercial sense and I think it is at least arguable that there can be a *receipt* of money even if there has been no physical receipt of it by the main contractor where his debt has been discharged by the employer exercising a right of set-off. No doubt that point will be argued out soon.

Table of Cases

Index